Quantum Physics

for Beginners

Unlocking the Secrets of Wave Theory, Quantum Computing, and Mechanics.

Understand the Fundamentals and How Everything Works in the Fascinating World of Quantum Physics.

Nathan Scott

QUANTUM PHYSICS FOR BEGINNERS

Text Copyright © [Nathan Scott]

All rights reserved. No part of this guide may be reproduced in any form without permission in writing from the publisher except in the case of brief quotations embodied in critical articles or reviews.

Legal & Disclaimer

The information contained in this book and its contents is not designed to replace or take the place of any form of medical or professional advice; and is not meant to replace the need for independent medical, financial, legal or other professional advice or services, as may be required. The content and information in this book has been provided for educational and entertainment purposes only.

The content and information contained in this book has been compiled from sources deemed reliable, and it is accurate to the best of the Author's knowledge, information and belief. However, the Author cannot guarantee its accuracy and validity and cannot be held liable for any errors and/or omissions. Further, changes are periodically made to this book as and when needed. Where appropriate and/or necessary, you must consult a professional (including but not limited to your doctor, attorney, financial advisor or such other professional advisor) before using any of the suggested remedies, techniques, or information in this book.

Upon using the contents and information contained in this book, you agree to hold harmless the Author from and against any damages, costs, and expenses, including any legal fees potentially resulting from the application of any of the information provided by this book. This disclaimer applies to any loss, damages or injury caused by the use and application, whether directly or indirectly, of any advice or information presented, whether for breach of contract, tort, negligence, personal injury, criminal intent, or under any other cause of action.

You agree to accept all risks of using the information presented inside this book.

You agree that by continuing to read this book, where appropriate and/or necessary, you shall consult a professional (including but not limited to your doctor, attorney, or financial advisor or such other advisor as needed) before using any of the suggested remedies, techniques, or information in this book.

Table of Contents

Introduction 6

Chapter 1: What is Quantum Physics? 8

    The Quantum Revolution: A Brief History 8

    The Quantum World: A Different Perspective 10

Chapter 2: Exploring Light and Wave Theory 15

    Understanding Light: An Enigma Unveiled 15

    The Speed of Light and Electromagnetic Waves 19

    Quantum Theory: Insights into the Subatomic World 23

    Revealing Quantum Particles: The Photoelectric and Compton Effects 26

Chapter 3: Embracing Wave-Particle Duality and Quantum Mechanics 31

    The Great Paradox: Wave-Particle Duality 31

    Schrödinger's Quantum Theory and the Unified Perspective ... 35

    New Atomic Model: Energy Levels and Quantum Jumps 39

    Unveiling Schrödinger's Cat and Quantum Superposition 42

    Embracing Uncertainty: The Heisenberg Uncertainty Principle 44

Chapter 4: Exploring Quantum Theories 49

Quantum Field Theory: Unifying Forces and Particles 49

Seeking Unity at the Cosmic Scale: Quantum Gravity 54

String Theory: Harmonies in a Multidimensional Universe 59

Quantum Entanglement: Spooky Connections and Teleportation 61

Chapter 5: Real-World Applications of Quantum Physics 67

Ultra-Precise Clocks: Timekeeping in the Quantum Realm 67

Quantum Key Distribution: Securing Communication 72

Quantum Computing: Unleashing the Power of Quantum Bits 77

Chapter 6: Reflecting on the Quantum Journey 82

Key Insights and Takeaways 82

Embracing the Mysteries: The Beauty of Quantum Physics 87

Bonus 1 92

Quantum Tunneling: Journeying Through Barriers. Unveiling the Phenomenon of Quantum Tunneling 92

Bonus 2 96

Physics and Superheroes Download: Unleashing the Science Behind Superpowers. Exploring the Physics of Superhero Abilities 96

Bonus 3 101

The Future of Quantum Physics: From Discoveries to Breakthroughs. Anticipating the Exciting Frontiers of Quantum Research..101

Introduction

Welcome to the fascinating world of quantum physics! In this book, "Quantum Physics for Beginners: Unlocking the Secrets of Wave Theory, Quantum Computing, and Mechanics," we will embark on a journey of discovery to unravel the mysteries of the quantum realm.

Have you ever wondered about the fundamental nature of reality? How does the universe function at its most fundamental level? Quantum physics provides us with the tools to explore these profound questions. From wave-particle duality to quantum entanglement, this field of science revolutionizes our understanding of the physical world and challenges our intuitions.

In this book, we will delve into the fundamental principles of quantum physics and demystify its complexities, making it accessible to beginners. We will explore the intriguing realm of wave theory, where light behaves both as a wave and a particle, and dive into the wonders of quantum mechanics, where particles can exist in multiple states simultaneously. Moreover, we will unlock the secrets of quantum computing and its potential to revolutionize technology.

The journey begins with an introduction to the basics of quantum physics, providing a historical context and presenting the unique perspective it offers. We will then shed light on the enigma of light, unraveling its dual nature and exploring its behavior through the lens of wave theory. From there, we will venture into the subatomic world, where quantum particles defy our everyday intuitions.

As we delve deeper into the quantum realm, we will encounter intriguing phenomena such as wave-particle duality and the famous Schrödinger's cat thought experiment. We will navigate through the uncertainty principle and its profound implications for our understanding of the physical world.

But our exploration doesn't end there. We will embark on a captivating journey through various quantum theories, including quantum field theory, quantum gravity, and string theory. These theories push the boundaries of our knowledge and offer tantalizing glimpses into the underlying fabric of the universe.

Finally, we will explore the real-world applications of quantum physics, such as ultra-precise clocks, secure communication through quantum key distribution, and the potential of quantum computing to solve complex problems.

By the end of this book, you will have a solid grasp of the fundamental concepts of quantum physics and a deeper appreciation for the captivating nature of this field. Whether you are a student, a science enthusiast, or simply curious about the mysteries of the quantum world, this book will guide you on an enlightening journey.

So, get ready to unlock the secrets of wave theory, quantum computing, and mechanics. Let's embark on this captivating adventure into the fascinating world of quantum physics!

Chapter 1: What is Quantum Physics?

The Quantum Revolution: A Brief History

In order to grasp the essence of quantum physics, it is essential to understand the revolutionary journey that has led to its development. The story of quantum physics is a captivating tale of scientific breakthroughs, paradigm shifts, and the gradual unveiling of the mysterious nature of the quantum world.

Our journey begins in the late 19th century, when physicists were faced with puzzling experimental observations that could not be explained by classical physics. These observations challenged the long-standing belief that the laws of classical physics were sufficient to explain the behavior of matter and energy. It was in this scientific climate that the foundations of quantum physics were laid.

One of the key figures in the early days of quantum physics was Max Planck. In 1900, Planck proposed a revolutionary idea to explain the phenomenon of blackbody radiation. He introduced the concept of quantization, suggesting that energy can only be absorbed or emitted in discrete, indivisible units called "quanta." This groundbreaking insight laid the foundation for a new branch of physics, known as quantum mechanics.

Building upon Planck's work, Albert Einstein further advanced the field with his groundbreaking explanation of the photoelectric effect. In 1905, Einstein proposed that light behaves not only as a wave but also as a particle, now known as a photon. This duality of light challenged the traditional understanding of the nature of light and paved the way for the development of quantum theory.

The true quantum revolution, however, came with the work of Danish physicist Niels Bohr and his colleagues. In the 1920s, Bohr introduced the concept of quantized energy levels in atoms, revolutionizing our understanding of atomic structure. Bohr's model, known as the Bohr model, provided a framework for explaining the discrete emission and absorption spectra observed in atomic systems. This model marked a significant departure from the classical understanding of atoms as small, solid spheres.

The birth of quantum mechanics as a comprehensive theory came with the contributions of pioneers such as Werner Heisenberg, Erwin Schrödinger, and Paul Dirac. Heisenberg formulated the uncertainty principle, which states that certain pairs of physical properties, such as position and momentum, cannot be precisely measured simultaneously. Schrödinger, on the other hand, developed the wave equation, a mathematical tool that describes the behavior of quantum particles as wave functions. Dirac's work on quantum mechanics paved the way for the development of quantum field theory.

The advent of quantum mechanics brought about a fundamental shift in our understanding of the physical world. It revealed that particles can exist in multiple states simultaneously and exhibit wave-like behavior. This wave-particle duality lies at the heart of quantum physics and challenges our intuition based on classical physics.

The journey of quantum physics is one of continual exploration and discovery. As our understanding of the quantum world deepens, new phenomena and theories continue to emerge. In the subsequent chapters of this book, we will explore the essential characteristics of quantum physics, including superposition and entanglement, and unravel the mysteries of wave-particle duality. We will also delve into the quantum world of light, investigating its dual nature and the implications of wave theory.

So, fasten your seat belts and get ready to embark on a captivating exploration of the quantum world. The journey ahead will be filled with fascinating concepts and mind-bending phenomena that will challenge your preconceived notions of reality. Let us now delve deeper into the captivating realm of quantum physics and unlock the secrets of wave theory, quantum computing, and mechanics.

The Quantum World: A Different Perspective

As we dive deeper into the realm of quantum physics, we begin to realize that the quantum world operates by its own set of rules and principles, vastly different from our everyday experiences. In this chapter, we will explore the fascinating aspects that make the quantum world so unique and intriguing.

One of the fundamental characteristics of the quantum world is superposition. Unlike classical physics, where objects exist in well-defined states, quantum particles can exist in multiple states simultaneously. It's as if they are in a dance of possibilities, with the ability to be in two or more states at once. This concept challenges our intuition and forces us to rethink our understanding of reality.

Imagine a subatomic particle, such as an electron, existing in a superposition of states. It can be in multiple locations, spinning in multiple directions, and even exhibiting contradictory properties, all at the same time. This mind-boggling behavior is at the core of quantum physics and has been confirmed through numerous experiments.

Another intriguing phenomenon in the quantum world is quantum entanglement. When two particles become entangled, their properties become deeply intertwined, regardless of the distance separating them. Changes made to one particle instantaneously affect the other, no matter how far apart they are. This phenomenon, famously referred to as "spooky action at a distance" by Einstein, challenged the concept of local realism and continues to be a subject of intense study and debate.

The concept of quantum entanglement has profound implications for communication, cryptography, and even the nature of space-time. Scientists are harnessing entanglement to develop quantum technologies such as quantum teleportation and quantum key distribution, which promise to revolutionize the fields of communication and data security.

In the quantum world, measurement takes on a whole new meaning. When we observe a quantum particle, its superposition collapses into a specific state. This collapse, known as wave function collapse, is a fundamental aspect of quantum physics. It demonstrates that the act of measurement has a profound impact on the observed system.

However, this poses a perplexing question: How can we reconcile the fact that a particle can exist in multiple states simultaneously, yet collapse into a single state upon measurement? This apparent contradiction is at the heart of the wave-particle duality dilemma, one of the most intriguing puzzles in quantum physics. It suggests that particles can exhibit both wave-like and particle-like behavior depending on how they are observed.

To navigate the quantum world, scientists use mathematical tools known as wave functions. These functions describe the probabilistic nature of quantum particles and allow us to make predictions about their behavior. The Schrödinger equation, developed by Erwin Schrödinger, is a cornerstone of quantum mechanics and provides a framework for understanding wave functions and their evolution over time.

The quantum world challenges our intuition and stretches the boundaries of our understanding. It forces us to think beyond the classical notions of cause and effect, determinism, and objective reality. The puzzles and paradoxes it presents are not merely intellectual curiosities but have practical implications for technology, computing, and our fundamental understanding of the universe.

In the subsequent chapters of this book, we will delve further into the intricacies of quantum physics, exploring the wave-particle duality, Schrödinger's quantum theory, and the new atomic model that emerged from quantum mechanics. We will witness the enigmatic thought experiment of Schrödinger's cat, uncover the mysteries of quantum superposition, and explore the limits of knowledge imposed by the Heisenberg uncertainty principle.

So, join me as we embark on a mind-expanding journey through the quantum world. It is a realm where the boundaries of reality blur, where particles exist in multiple states, and where the very fabric of our understanding is challenged. Together, we will unlock the secrets of wave theory, quantum computing, and mechanics, and gain a profound appreciation for the wonders of quantum physics.

As we venture deeper into this realm, it's important to keep an open mind and embrace the unique perspective that the quantum world offers. The principles and phenomena we will encounter may seem strange and counterintuitive at first, but they are firmly grounded in rigorous scientific research and experimentation.

Throughout this book, we will rely on clear explanations and relatable examples to demystify complex concepts. I will guide you step by step, ensuring that you grasp the fundamental principles of quantum physics and their practical applications. We will explore the historical milestones that led to the development of quantum theory, from Max Planck's groundbreaking discovery of quantized energy to Albert Einstein's debates on the nature of light.

By understanding the basics of quantum physics, you will gain a solid foundation for comprehending the astonishing advancements and breakthroughs that have emerged in recent decades. We will explore the mysterious wave-particle duality and delve into the famous double-slit experiment, which revealed the puzzling nature of quantum particles.

Moreover, we will delve into the revolutionary ideas put forth by Erwin Schrödinger, whose quantum wave equation has transformed our understanding of the microscopic world. Through his theories, we will unravel the enigmatic concept of superposition, where particles can exist in a multitude of states simultaneously. We will also encounter Schrödinger's infamous thought experiment involving a cat in a paradoxical state of being simultaneously alive and dead.

In our exploration of quantum physics, we will also examine the profound implications and applications of these principles. From quantum computing, which promises to revolutionize computational power, to the cutting-edge field of quantum cryptography, where the security of information is protected through the intricacies of quantum mechanics.

Throughout this journey, I encourage you to ask questions, ponder the mysteries of the quantum world, and embrace the excitement of discovery. Quantum physics has the potential to transform our understanding of the universe and reshape the technological landscape. By unraveling its secrets, we gain not only a deeper understanding of the fundamental nature of reality but also a glimpse into the incredible possibilities that lie ahead.

So, let us embark on this captivating expedition into the fascinating world of quantum physics. Together, we will unlock the secrets of wave theory, delve into the mind-bending world of quantum computing, and explore the intricate mechanics that govern the subatomic realm. Brace yourself for an adventure that will challenge your preconceptions, expand your horizons, and ignite your curiosity. Get ready to witness the marvels of the quantum world and grasp the fundamental principles that underpin its profound influence on our understanding of the universe.

Welcome to the enthralling realm of quantum physics!

Chapter 2: Exploring Light and Wave Theory

Understanding Light: An Enigma Unveiled

Light, the seemingly intangible phenomenon that surrounds us, has captivated the human mind for centuries. It illuminates our world, reveals its vibrant colors, and enables us to perceive the wonders of our surroundings. But what exactly is light? How does it behave? And what secrets does it hold within the realm of quantum physics?

In this chapter, we will embark on a journey to unravel the mysteries of light and delve into the fascinating field of wave theory. We will explore the enigmatic nature of light, its dual behavior as both a wave and a particle, and the profound impact it has had on the development of quantum physics.

To understand light, we must first recognize its historical significance. Throughout the ages, numerous scholars and scientists have contributed to our understanding of this phenomenon. From ancient Greek philosophers pondering the nature of vision to groundbreaking experiments conducted by pioneers like Isaac Newton and Thomas Young, the study of light has shaped our perception of the world.

Let us start by examining the nature of light itself. For centuries, the prevailing belief was that light traveled in a straight line, much like a stream of particles. However, the wave theory of light challenged this notion and proposed that light behaves as a wave, propagating through space. This revolutionary idea was championed by eminent scientists such as Christiaan Huygens and Augustin-Jean Fresnel.

As we delve into the depths of wave theory, we encounter a fundamental concept: the wavelength. The wavelength of light determines its color and is a crucial characteristic of any wave. Through intricate experiments, we have come to understand that different colors correspond to different wavelengths, with red having a longer wavelength than blue, for example. This discovery laid the foundation for the study of light's diverse properties and interactions with matter.

The wave theory of light also offers an explanation for phenomena such as diffraction and interference. When light encounters an obstacle or passes through a narrow slit, it undergoes bending and spreading, creating intricate patterns of light and dark regions. This phenomenon, known as diffraction, provides evidence for light's wave-like nature.

Furthermore, when two or more light waves converge or interfere with each other, they can reinforce or cancel each other out, resulting in a phenomenon called interference. This captivating interplay of light waves has been observed in various experiments, highlighting the wave-like behavior of light and deepening our understanding of its nature.

While wave theory successfully explains many properties of light, it faced a significant challenge in the late 19th and early 20th centuries. The discoveries of phenomena such as the photoelectric effect and the emission of light by atoms challenged the wave theory's ability to fully account for the observed phenomena. These breakthroughs paved the way for the birth of quantum physics and the realization that light also exhibits particle-like behavior.

This realization set the stage for the development of quantum mechanics and the exploration of the wave-particle duality. In the early 20th century, physicists such as Max Planck and Albert Einstein proposed that light can behave both as a wave and as discrete packets of energy known as photons. This revolutionary concept reconciled the seemingly contradictory observations and laid the groundwork for a deeper understanding of light and its interaction with matter.

The duality of light is perhaps best exemplified by the famous double-slit experiment. When a beam of light passes through two closely spaced slits, it creates an interference pattern on a screen behind them, as if it were a wave. However, when the intensity of the light is reduced to the point where only one photon passes through at a time, the individual photons behave as particles, creating distinct impact points on the screen.

This experiment shattered the classical notion of particles as solid, distinct entities and emphasized the wave-particle duality inherent in quantum physics. It demonstrated that light possesses both wave-like and particle-like characteristics, challenging our intuitive understanding of the nature of reality.

The wave-particle duality of light is not limited to photons alone. It extends to other subatomic particles as well, such as electrons and protons. This profound realization opened the doors to the exploration of quantum mechanics, a branch of physics that seeks to understand the behavior of particles at the smallest scales.

Quantum mechanics provides us with a mathematical framework to describe the behavior of particles in the quantum realm. It introduces concepts such as wave functions, which describe the probability distribution of a particle's position or properties. Through mathematical equations, such as Schrödinger's equation, we can calculate the probabilities of various outcomes when observing particles in quantum systems.

As we delve further into the quantum world, we encounter fascinating phenomena that defy our classical intuitions. One such phenomenon is quantum entanglement, a phenomenon where two particles become linked in such a way that the state of one particle instantly affects the state of the other, regardless of the distance between them. This perplexing behavior, famously referred to by Albert Einstein as "spooky action at a distance," challenges our notions of causality and necessitates a new understanding of the interconnectedness of particles.

Understanding light and its wave-particle duality is crucial to grasping the fundamentals of quantum physics. It serves as a gateway to exploring the intricate and mysterious world of quantum phenomena. By embracing the notion that light can behave as both a wave and a particle, we unlock the secrets of quantum physics and gain insight into the behavior of particles at the smallest scales.

In the following chapters, we will continue our exploration of quantum physics, delving into topics such as the uncertainty principle, the Schrödinger's cat thought experiment, and the emerging field of quantum computing. We will uncover the fascinating implications of quantum theory and its applications in various fields, ranging from advanced timekeeping to secure communication and computing.

As we embark on this journey, let us embrace the wonders of the quantum world and expand our understanding of the fundamental building blocks of the universe. Quantum physics invites us to question our perceptions, challenge our preconceived notions, and dive into a realm where the ordinary rules of classical physics no longer hold sway.

So, join me as we unravel the mysteries of wave theory, quantum computing, and mechanics. Together, we will unlock the secrets of quantum physics and gain a deeper appreciation for the remarkable world that lies beyond our everyday experiences. Let the adventure begin!

The Speed of Light and Electromagnetic Waves

In our quest to unravel the mysteries of light and delve into the captivating realm of electromagnetic waves, we embark on a journey that takes us deeper into the fundamental nature of these phenomena. By understanding the speed at which light travels and exploring the intricate properties of electromagnetic waves, we gain valuable insights into the underlying fabric of the universe.

At the heart of our exploration lies the remarkable constant known as the speed of light. Denoted by the symbol "c," this universal constant holds a pivotal role in our understanding of the cosmos. In the vacuum of space, light travels at an astonishing velocity of approximately 299,792,458 meters per second (or about 186,282 miles per second). This unchanging speed acts as a cosmic speed limit, establishing a fundamental limit on how quickly information can travel through the universe.

The concept of the speed of light revolutionized our understanding of space, time, and causality. It played a pivotal role in the development of Albert Einstein's theory of relativity, which introduced profound insights into the nature of space and time. According to the theory of special relativity, the speed of light is the same for all observers, regardless of their relative motion. This principle led to revolutionary concepts such as time dilation and length contraction, challenging our intuitive notions of how the universe operates.

Moving forward, we delve into the captivating realm of electromagnetic waves, which form the foundation of our understanding of light and its interactions with matter. Electromagnetic waves encompass a vast spectrum of energy, ranging from the longest radio waves to the shortest gamma rays. They are composed of oscillating electric and magnetic fields, perpetually intertwined in an elegant dance dictated by Maxwell's equations—fundamental equations in classical electromagnetism.

To comprehend the behavior of electromagnetic waves, we turn to the principles of wave theory. According to this theory, light can be described as a transverse wave, characterized by oscillations perpendicular to the direction of its propagation. Imagine a pebble dropped into a calm pond, witnessing the ripples radiating outward as the wave propagates through space.

Two fundamental parameters dictate the properties of electromagnetic waves: wavelength and frequency. The wavelength represents the distance between consecutive crests or troughs of a wave, while the frequency indicates the number of oscillations occurring per second. These parameters are inversely related—a longer wavelength corresponds to a lower frequency, and a shorter wavelength corresponds to a higher frequency.

QUANTUM PHYSICS FOR BEGINNERS

The electromagnetic spectrum encompasses a vast range of wavelengths and frequencies, each corresponding to a unique type of electromagnetic wave. At the longest wavelength end, we encounter radio waves, which have frequencies ranging from a few kilohertz to gigahertz. These waves find applications in various technologies, including telecommunications, broadcasting, and radar systems.

As we move along the spectrum, we encounter microwaves, which are utilized in applications such as cooking, wireless communications, and satellite communications. Further up the spectrum, we find infrared radiation, which is invisible to the human eye but is responsible for phenomena such as thermal imaging and remote controls.

Continuing our journey, we arrive at the visible light range, the narrow band of the spectrum that our eyes perceive. This range spans from approximately 400 to 700 nanometers in wavelength and encompasses the colors of the rainbow. Our understanding of the properties of visible light paved the way for significant advancements in fields such as optics, photography, and display technologies.

Beyond visible light, we encounter the fascinating realms of ultraviolet radiation, X-rays, and gamma rays. Ultraviolet radiation, with its shorter wavelengths, plays a vital role in processes such as sunscreen protection, fluorescence, and sterilization. X-rays find applications in medical imaging, material testing, and airport security, allowing us to peer into the inner structure of objects. Lastly, at the highest end of the spectrum, we encounter gamma rays, which possess the shortest wavelengths and highest energies. These powerful rays are generated by celestial events such as supernovae and black holes, and their study provides us with valuable insights into the extreme phenomena occurring in the cosmos.

Understanding the properties of electromagnetic waves not only allows us to appreciate the diverse range of phenomena they encompass but also enables us to harness their power for various applications. From the transmission of information through radio waves to the precise imaging capabilities of X-rays, our technological advancements heavily rely on our understanding of electromagnetic waves.

Moreover, the behavior of light and electromagnetic waves is not limited to their propagation through empty space. When encountering different media, such as air, water, or glass, light can undergo phenomena such as reflection, refraction, and diffraction. These interactions play a crucial role in everyday experiences, from the formation of rainbows to the bending of light in lenses and prisms.

As we conclude our exploration of light and wave theory, we have gained a deeper appreciation for the profound influence of electromagnetic waves in our understanding of the universe and the technological advancements they have facilitated. The concepts of wavelength, frequency, and the speed of light provide us with a powerful toolkit for comprehending the nature of light and its interactions with matter.

Through this journey, we have uncovered the fundamental principles that underpin the behavior of light, paving the way for our exploration of quantum physics and its enigmatic phenomena. In the following chapters, we will delve further into the quantum world, where the boundaries of classical physics blur and give rise to a fascinating tapestry of particles, waves, and uncertainty.

Quantum Theory: Insights into the Subatomic World

In our exploration of light and wave theory, we have encountered the fascinating properties of electromagnetic waves and their role in understanding the nature of light. Now, we venture deeper into the subatomic world, where the rules of classical physics no longer hold sway. Welcome to the realm of quantum theory, a groundbreaking framework that has revolutionized our understanding of the universe.

Quantum theory, also known as quantum mechanics, provides a unique perspective on the behavior of particles at the atomic and subatomic levels. It describes a realm where particles, such as electrons and photons, exhibit both wave-like and particle-like properties, challenging our classical notions of duality. To grasp the essence of quantum theory, we must delve into its foundational concepts and the experiments that have shaped our understanding.

At the heart of quantum theory lies the concept of quantization. According to this principle, certain physical quantities, such as energy and angular momentum, can only take discrete values, or "quanta." This departure from the continuous spectrum of classical physics has profound implications for the behavior of particles and the interactions between them.

One of the key features of quantum theory is the wave-particle duality. This concept states that particles can exhibit wave-like properties, such as diffraction and interference, while waves can also manifest particle-like behavior, as seen in the discrete nature of energy emission and absorption. This duality was famously illustrated by the double-slit experiment, where particles, when sent through a barrier with two slits, create an interference pattern characteristic of waves. This experiment highlighted the puzzling nature of quantum entities and paved the way for further investigations into their behavior.

Quantum theory also introduces the notion of superposition, which is the ability of particles to exist in multiple states simultaneously. In other words, a particle can be in a combination of different states until it is measured or observed, at which point it "collapses" into a single state. This idea challenges our intuition but has been experimentally confirmed through various studies.

Another intriguing aspect of quantum theory is quantum entanglement. This phenomenon occurs when two or more particles become intertwined in such a way that their states are intimately linked. The intriguing feature of entanglement is that the states of the particles remain connected regardless of the distance between them, suggesting a type of instantaneous correlation that defies classical notions of causality.

Quantum theory also introduces the concept of the wave function, which mathematically describes the probabilistic nature of particles. The wave function allows us to calculate the probability of finding a particle in a particular state and predicts the statistical outcomes of measurements. The behavior of particles at the quantum level is inherently uncertain, and the wave function provides a framework for understanding and quantifying this uncertainty.

Through a series of groundbreaking experiments, such as the Stern-Gerlach experiment and the famous thought experiment known as Schrödinger's cat, scientists have continuously probed the mysteries of the quantum world. These experiments have challenged our understanding of reality and unveiled the peculiar nature of quantum phenomena.

As we explore quantum theory, we begin to appreciate its fundamental role in shaping our understanding of the subatomic world. Its concepts and principles provide a powerful framework for comprehending the behavior of particles and the interactions that govern their existence. Quantum theory forms the basis for numerous technological advancements, including quantum computing, quantum cryptography, and quantum teleportation.

In the next chapters, we will further unravel the intricacies of quantum theory and its profound implications for the understanding of the universe. We will delve into the quantum nature of particles, the uncertainty principle, and the exciting possibilities of quantum entanglement. Prepare to be captivated by the extraordinary world of quantum physics, where the boundaries of our classical intuition are pushed to their limits, and where new frontiers of knowledge await our exploration.

In conclusion, the study of quantum theory is a journey into the mysterious and awe-inspiring subatomic realm. It challenges our classical understanding of physics and offers profound insights into the fundamental nature of reality. From the wave-particle duality to superposition and entanglement, quantum theory provides a rich tapestry of concepts that have revolutionized our scientific understanding.

As we continue our exploration, we will delve into the applications of quantum theory, including its role in the development of quantum technologies. From ultra-precise clocks to secure communication and the promises of quantum computing, these applications demonstrate the tangible impact of quantum physics on our daily lives and the potential for transformative advancements in various fields.

Join us as we unravel the mysteries of the quantum world and embark on an extraordinary journey of discovery. Through this book, we aim to demystify the complex principles of quantum physics, making them accessible to beginners while maintaining scientific rigor and accuracy.

So, fasten your seatbelts and prepare to dive into the fascinating world of quantum physics. In the following chapters, we will explore the quantum properties of matter, the enigmatic wave-particle duality, the fundamental principles of quantum mechanics, and the exciting frontiers of quantum research. By the end of this journey, you will gain a deeper appreciation for the intricate nature of the quantum realm and its impact on our understanding of the universe.

Revealing Quantum Particles: The Photoelectric and Compton Effects

In our exploration of light and its connection to quantum physics, we delve into two remarkable phenomena that provide crucial insights into the particle-like behavior of light and the existence of quantum particles: the photoelectric effect and the Compton effect.

The Photoelectric Effect: Unveiling Quantum Particles

The photoelectric effect, first observed by Heinrich Hertz in 1887 and later explained by Albert Einstein in 1905, revolutionized our understanding of the nature of light and the behavior of electrons.

This experiment involves the emission of electrons from a material when it is illuminated with light of a certain frequency or energy.

Classical wave theory predicted that increasing the intensity of the light would eventually cause electrons to be emitted, regardless of the frequency. However, experimental observations presented a puzzling contradiction to this expectation. Einstein's groundbreaking explanation introduced a new perspective based on the quantization of energy.

According to Einstein's interpretation, light consists of discrete packets of energy called photons. These photons carry energy proportional to their frequency, and when a photon interacts with a material, it can transfer its energy to an electron, resulting in the ejection of the electron from the material. However, for this emission to occur, the energy of the photon must surpass a specific threshold known as the work function of the material.

The photoelectric effect demonstrated the particle-like behavior of light and provided compelling evidence for the existence of quantized energy levels in atoms. It played a pivotal role in establishing the concept of the photon as a fundamental particle of light and paved the way for the development of quantum theory.

Moreover, the photoelectric effect revealed additional intriguing aspects. The kinetic energy of the emitted electrons was found to depend on the frequency of the incident light rather than its intensity. This discovery challenged classical wave theory's prediction that the energy transferred to the electrons should increase with the intensity of the light. Einstein's explanation addressed this discrepancy by postulating that the energy of a photon is directly proportional to its frequency, not its intensity.

The photoelectric effect also introduced the concept of the work function, which represents the minimum energy required to liberate an electron from the material. Different materials possess unique work functions, resulting in varying thresholds for the emission of electrons. This observation further reinforced the particle-like nature of light and provided valuable insights into the quantized behavior of electrons.

The Compton Effect: Revealing the Particle Nature of Light

The Compton effect, discovered by Arthur H. Compton in 1923, further elucidates the particle-like nature of light and its interaction with matter. This phenomenon involves the scattering of X-rays or gamma rays by electrons.

Classical wave theory predicted that the wavelength of scattered radiation should solely depend on the angle of scattering. However, Compton's experiments revealed a different outcome. He observed that the scattered radiation exhibited a longer wavelength than the incident radiation, and the change in wavelength was directly related to the scattering angle.

To explain this phenomenon, Compton proposed a model in which photons behave as particles that collide with electrons. In this model, when a photon collides with an electron, it transfers a portion of its energy and momentum to the electron, causing the electron to recoil. Consequently, the scattered photon undergoes a change in wavelength, corresponding to a decrease in energy.

The Compton effect provided compelling evidence for the particle-like nature of light and confirmed the existence of photons as discrete entities possessing both energy and momentum. It demonstrated the necessity of treating light as both a wave and a particle and reinforced the fundamental principles of quantum theory.

Furthermore, the Compton effect revealed additional insights into the relationship between wavelength, energy, and momentum. The change in wavelength observed in the scattered radiation can be quantitatively described using Compton's formula, which relates the change in wavelength to the scattering angle and the mass of the electron. This formula provided experimental confirmation of the wave-particle duality of light and its interaction with matter.

The Compton effect not only confirmed the particle-like behavior of light but also provided a means to directly measure the wavelength and momentum of photons. By analyzing the scattered radiation, scientists were able to determine the wavelength shift and subsequently calculate the momentum transfer between the photon and the electron. This groundbreaking experiment provided tangible evidence for the existence of quantized particles of light and their interactions at the atomic level.

Moreover, the Compton effect played a crucial role in the development of X-ray crystallography, a powerful technique for studying the atomic structure of materials. By analyzing the scattered X-rays from a crystal lattice, scientists can extract valuable information about the arrangement of atoms, providing insights into the fundamental properties and behaviors of matter.

The combination of the photoelectric effect and the Compton effect unveiled the dual nature of light and its intimate connection to quantum particles. These experiments shattered the classical wave theory's limitations and laid the foundation for the birth of quantum mechanics, a revolutionary framework that profoundly transformed our understanding of the microscopic world.

In summary, the photoelectric effect and the Compton effect provided groundbreaking experimental evidence for the particle-like behavior of light and its interaction with matter. They revealed the quantized nature of energy, the existence of photons as discrete packets of energy and momentum, and the wave-particle duality inherent in quantum physics. These profound insights paved the way for the development of quantum theory, which continues to unravel the mysteries of the subatomic realm.

Chapter 3: Embracing Wave-Particle Duality and Quantum Mechanics

The Great Paradox: Wave-Particle Duality

In the fascinating world of quantum physics, one of the most perplexing phenomena is wave-particle duality. It challenges our classical understanding of nature by revealing that particles can exhibit both wave-like and particle-like behavior simultaneously. This profound concept, first discovered through various experimental observations and theoretical developments, has revolutionized our perception of the fundamental building blocks of the universe.

Wave-particle duality introduces a profound paradox: How can an object be both a wave and a particle at the same time? To comprehend this paradox, we need to explore the key experiments and theories that have unraveled this mystery.

One of the pivotal experiments that revealed wave-particle duality is the famous double-slit experiment. Imagine a barrier with two narrow slits, and a beam of particles, such as electrons or photons, directed towards it. Classical intuition would suggest that the particles should behave as individual entities and create two separate bands of particles on the screen behind the slits. However, what the experiment demonstrates is quite astonishing.

When the particles are fired one by one towards the double-slit barrier, an interference pattern emerges on the screen, similar to the pattern observed when light passes through two closely spaced slits and creates an interference pattern of bright and dark bands. This interference pattern indicates that the particles are interfering with themselves, as if they are behaving like waves. Somehow, each individual particle is able to go through both slits and interfere with itself to create this pattern.

The double-slit experiment not only confirmed the wave-like behavior of particles but also raised intriguing questions about their nature. How does a single particle simultaneously pass through both slits and interfere with itself? This puzzling behavior challenged the classical notion of particles as localized entities with definite trajectories.

To address this paradox, the development of quantum mechanics provided a powerful theoretical framework. In particular, Erwin Schrödinger's wave equation, formulated in 1926, revolutionized our understanding of wave-particle duality. Schrödinger's equation describes the wave-like behavior of particles through a mathematical function called the wave function. This function represents the probability distribution of finding a particle at different positions in space.

The wave function allows us to calculate the probability of a particle exhibiting certain properties, such as its position or momentum. Importantly, it also captures the wave-like behavior of particles, enabling us to understand phenomena like interference and diffraction. The square of the wave function, known as the probability density, provides the likelihood of finding a particle at a specific location.

Schrödinger's equation and the concept of the wave function revolutionized quantum mechanics by providing a mathematical framework to describe the behavior of particles in a wave-like manner. It unified the seemingly contradictory wave and particle aspects, giving rise to a new understanding of the subatomic world.

Furthermore, the wave-particle duality of light was further elucidated by the groundbreaking work of Louis de Broglie, who proposed that particles, not just photons, could also exhibit wave-like properties. De Broglie's hypothesis introduced the concept of matter waves, suggesting that all particles, including electrons and atoms, possess wave-like characteristics. This profound insight laid the foundation for wave mechanics and expanded the wave-particle duality concept beyond light.

The concept of wave-particle duality has far-reaching implications, extending beyond the behavior of particles in experiments. It underlies the fundamental principles of quantum mechanics, shaping our understanding of the microscopic world. The ability of particles to exhibit both wave-like and particle-like characteristics provides a deeper understanding of phenomena such as energy quantization, electron orbitals, and the probabilistic nature of quantum events.

By embracing wave-particle duality and delving into the depths of quantum mechanics, we uncover a rich tapestry of interconnected concepts that tie together the intricate workings of the quantum world. The phenomenon of wave-particle duality serves as a gateway to explore the fundamental principles that govern the behavior of subatomic particles.

As we delve deeper into the realm of quantum mechanics, we encounter the concept of superposition. Superposition states that particles can exist in multiple states simultaneously, akin to a wave existing in multiple places at once. This concept allows particles to occupy multiple energy levels or spin orientations, leading to the intriguing phenomenon of quantum superposition.

Another fundamental aspect of wave-particle duality is the concept of quantum entanglement. When particles interact in a way that their quantum states become correlated, they become entangled. This entanglement can persist even when the particles are separated by vast distances, defying classical notions of locality. The entangled particles, regardless of the spatial separation, exhibit a mysterious connection, where the state of one particle instantaneously affects the state of the other, regardless of the distance between them.

Wave-particle duality and its associated principles are at the heart of quantum mechanics, guiding our understanding of phenomena such as the wave-like behavior of electrons in atoms, the behavior of particles in quantum fields, and the manipulation of quantum information in quantum computing.

The exploration of wave-particle duality has not only deepened our comprehension of the quantum world but has also led to practical applications. For instance, the understanding of wave-like properties of particles has paved the way for electron microscopy, enabling scientists to observe and manipulate matter at atomic scales. Additionally, the field of quantum information and quantum computing relies on the principles of wave-particle duality to harness the power of quantum states for computational purposes.

The journey into the realm of wave-particle duality and quantum mechanics is an ongoing endeavor, filled with mysteries waiting to be unraveled. It challenges our intuition and pushes the boundaries of our understanding. However, by embracing the paradoxes and delving into the intricate nature of the quantum world, we gain insights into the fundamental nature of reality itself.

In this chapter, we have explored the enigmatic phenomenon of wave-particle duality and its implications in quantum mechanics. We have examined the double-slit experiment, Schrödinger's wave equation, the concept of the wave function, and the far-reaching impact of wave-particle duality on our understanding of the microscopic world. By peering into the depths of quantum mechanics, we have caught a glimpse of the intricate interplay between waves and particles, unraveling the mysteries of the subatomic realm.

Schrödinger's Quantum Theory and the Unified Perspective

In our exploration of quantum mechanics, we delve into one of its most profound and influential theories: Schrödinger's quantum theory. Introduced by the Austrian physicist Erwin Schrödinger in the 1920s, this theory revolutionized our understanding of the quantum world, providing us with a unified perspective that reconciles the wave-particle duality and reveals the intricate interplay between waves and particles.

At the heart of Schrödinger's theory lies the wave equation, also known as the Schrödinger equation. This fundamental equation in quantum mechanics describes the behavior of quantum systems, providing us with a mathematical framework to calculate their probabilities and make predictions about their properties. The wave equation embodies the wave-like nature of particles and allows us to analyze their evolution over time.

The concept of the wave function is central to Schrödinger's quantum theory. Represented by the Greek letter Ψ (psi), the wave function encapsulates the complete information about a quantum system. It describes the probability distribution of finding a particle in a particular state or location. By squaring the wave function, $|\Psi|^2$, we obtain the probability density of locating the particle, enabling us to make statistical predictions.

One of the remarkable features of Schrödinger's theory is its ability to reconcile the dual nature of particles. It allows us to perceive particles as both waves and discrete entities, depending on the experimental conditions. This wave-particle duality is a fundamental aspect of the quantum world, and Schrödinger's equation allows us to calculate the probability distribution while retaining the wave-like characteristics.

Schrödinger's quantum theory provides a powerful framework for understanding and predicting the behavior of quantum systems across various scales. It enables us to determine the energy levels of atoms, decipher the behavior of molecules, and explore the interactions between particles. The theory has far-reaching implications, with applications ranging from the development of advanced materials to the design of quantum algorithms for information processing.

By embracing Schrödinger's unified perspective, we gain insight into the interconnected nature of the quantum realm. It bridges the gap between classical physics and the enigmatic world of the microscopic, offering a cohesive understanding of the fundamental principles at play. Schrödinger's theory forms the foundation upon which other influential theories, such as Heisenberg's matrix mechanics and Dirac's formulation of quantum mechanics, have been built, further expanding our comprehension of the quantum world.

In this chapter, we have explored Schrödinger's quantum theory and its significance in our quest to unravel the mysteries of the quantum world. We have delved into the intricacies of the wave equation, the concept of the wave function, and the profound notion of wave-particle complementarity. Through the lens of Schrödinger's unified perspective, we have gained a deeper appreciation for the rich and complex nature of quantum phenomena.

Schrödinger's quantum theory invites us to contemplate the profound implications of wave-particle duality and challenges our conventional understanding of reality. It compels us to confront the paradoxical nature of the quantum world, where particles can exist in multiple states simultaneously, and their properties become probabilistic in nature. This inherent uncertainty, captured elegantly by Schrödinger's equation, has sparked intense debates and philosophical ponderings among scientists and thinkers for decades.

As we journey deeper into Schrödinger's quantum theory, we encounter intriguing concepts such as superposition and quantum entanglement. Superposition allows particles to exist in multiple states simultaneously, presenting a departure from classical physics where objects are confined to definite states. Quantum entanglement, on the other hand, reveals the mysterious phenomenon of Quantum entanglement, on the other hand, reveals the mysterious phenomenon of non-local correlations between particles. When two or more particles become entangled, their quantum states become inseparably linked, regardless of the distance between them. This means that measuring the state of one entangled particle instantaneously affects the state of its entangled counterparts, even if they are light-years apart. This bizarre characteristic of entanglement challenges our classical notions of locality and raises profound questions about the nature of reality and the fundamental interconnectedness of the quantum world.

Schrödinger's quantum theory also provides a framework for understanding the quantum behavior of complex systems. In the realm of quantum chemistry, it allows us to investigate the electronic structure of atoms and molecules, predicting their energy levels, bonding patterns, and spectroscopic properties. The applications of quantum theory extend to fields such as materials science, where it enables the design and characterization of novel materials with tailored properties and functionalities. Moreover, Schrödinger's equation serves as the basis for computational methods that simulate quantum systems, opening doors to the exploration of quantum algorithms and the potential of quantum computing.

By studying Schrödinger's quantum theory, we gain a deeper appreciation for the intricate and counterintuitive nature of the quantum world. It challenges our classical intuitions and invites us to embrace a new way of thinking, where probabilities, superpositions, and entanglement become fundamental aspects of reality. The beauty of Schrödinger's theory lies in its ability to provide a mathematical framework that accurately describes the behavior of quantum systems, allowing us to make precise predictions and explore the frontiers of scientific discovery.

As we delve further into the fascinating world of quantum mechanics, we will explore other influential theories and concepts that expand our understanding of the quantum realm. From Heisenberg's uncertainty principle to the principles of quantum measurement and quantum field theory, each facet adds to the intricate tapestry of quantum physics. Our journey through the chapters of this book will take us on a comprehensive exploration of the principles, phenomena, and applications of quantum physics, unlocking the secrets of the subatomic realm.

In this chapter, we have unraveled the significance of Schrödinger's quantum theory and its role in shaping our understanding of wave-particle duality and the unified perspective of quantum mechanics. We have examined the wave equation, the wave function, and the profound concept of entanglement. These foundational elements pave the way for the subsequent chapters, where we will continue to deepen our knowledge and explore the captivating world of quantum physics.

New Atomic Model: Energy Levels and Quantum Jumps

In our fascinating journey into the realm of quantum mechanics, we encounter a revolutionary atomic model that challenges the classical understanding put forth by Bohr. This groundbreaking model, known as the wave-mechanical model or the quantum mechanical model, delves deeper into the behavior of electrons within atoms and introduces us to the intriguing concepts of energy levels and quantum jumps.

Within the wave-mechanical model, we move beyond the notion of electrons confined to specific orbits. Instead, electrons are described by complex mathematical constructs called wave functions, which represent their probability distributions. These wave functions offer insights into the likelihood of finding an electron at a particular location within the atom. By examining the square of the wave function, known as the probability density, we gain a deeper understanding of the probability of locating the electron at a specific point.

The concept of energy levels emerges as a result of the quantization of energy within the quantum mechanical model. Electrons in atoms can only occupy discrete energy levels, each with its own unique energy value. These energy levels are often visualized as electron shells or orbitals encircling the atomic nucleus. At the lowest energy level, known as the ground state, the electron resides in the region closest to the nucleus. As we progress to higher energy levels, the electron's distance from the nucleus increases, leading to a more intricate electronic structure.

Quantum jumps, also referred to as electron transitions or quantum leaps, occur when an electron undergoes a transition from one energy level to another. These transitions involve the absorption or emission of energy in precise, quantized amounts known as quanta or photons. When an electron absorbs energy, it ascends to a higher energy level, while the release of energy causes the electron to descend to a lower energy level. The energy associated with the absorbed or emitted photon corresponds precisely to the difference in energy between the initial and final energy levels.

The probabilities of these quantum jumps are determined by the wave functions of the electrons. These mathematical constructs provide crucial information about the likelihood of finding an electron in a particular energy state. It is essential to note that, in the quantum realm, electrons can exist in a superposition of energy states, meaning they can simultaneously occupy multiple energy levels. Consequently, the probability of undergoing transitions varies depending on the specific conditions and interactions.

The study of energy levels and quantum jumps plays a pivotal role in understanding diverse phenomena, such as atomic spectroscopy and the absorption and emission of light by atoms. Spectroscopy enables us to probe the energy levels of atoms by carefully examining the specific wavelengths of light absorbed or emitted during electron transitions. This invaluable knowledge finds practical applications in numerous fields, including astronomy, where spectroscopic analysis of light from distant celestial objects provides profound insights into their composition and properties.

Moreover, a comprehensive understanding of energy levels and quantum jumps paves the way for groundbreaking advancements in technologies such as lasers and quantum electronics. By precisely manipulating electron transitions, we can harness the unique properties of quantum systems for information processing, secure communication, and highly accurate measurements. These advancements hold tremendous potential for revolutionizing various aspects of our daily lives.

By embracing the new atomic model and diving deep into the concepts of energy levels and quantum jumps, we expand our understanding of the quantum world and gain profound insights into the behavior of atoms and their constituents. It is through this model that we unravel the mysteries of atomic structure and lay the foundation for further exploration into the captivating realms of quantum physics.

Unveiling Schrödinger's Cat and Quantum Superposition

In our exploration of quantum mechanics, we encounter a thought experiment that has captivated the imagination of physicists and non-physicists alike: Schrödinger's Cat. Conceived by the brilliant physicist Erwin Schrödinger in 1935, this intriguing paradox pushes the boundaries of our understanding of reality at the quantum level, highlighting the mind-boggling phenomenon known as quantum superposition.

Imagine a scenario where a cat is placed inside a sealed box, along with a radioactive substance and a mechanism that is triggered by the decay of a single atom. The fate of the cat, whether it is alive or dead, hinges on the unpredictable behavior of this atomic decay. According to the principles of quantum mechanics, until the box is opened and observed, the cat exists in a state of superposition—a bizarre state of being both alive and dead simultaneously.

Schrödinger's Cat serves as a gateway to delve deeper into the concept of quantum superposition. It illustrates the peculiar nature of quantum reality, where particles and systems can exist in a combination of multiple states at once. This notion challenges our classical intuitions and prompts us to explore the dualistic nature of matter and energy.

Quantum superposition is not limited to microscopic particles; it can extend to larger systems, such as molecules and even macroscopic objects. The wave-particle duality that we discussed earlier plays a crucial role in understanding superposition. Just as particles can exhibit both wave-like and particle-like properties, they can also exist in superposition, occupying multiple states simultaneously. This fundamental characteristic is at the heart of various quantum phenomena and forms the basis of modern quantum technologies.

The implications of superposition go beyond Schrödinger's Cat and have profound consequences for our understanding of the quantum world. The Copenhagen interpretation, one of the prominent interpretations of quantum mechanics, suggests that the act of observation causes the collapse of the wave function, leading to the manifestation of a definite state. In the case of Schrödinger's Cat, opening the box and observing the cat's condition forces the superposition to resolve into either a live or dead state.

Superposition also paves the way for another extraordinary phenomenon: quantum entanglement. Entanglement occurs when two or more particles become intricately correlated, such that the state of one particle is instantly connected to the state of the others, regardless of their physical separation. This concept, famously referred to as "spooky action at a distance" by Albert Einstein, is a remarkable manifestation of quantum mechanics.

Entanglement gives rise to the creation of quantum systems with properties that cannot be independently described. When particles become entangled, their states become inseparably intertwined, even if they are located at opposite ends of the universe. This non-local correlation has profound implications for quantum information processing, quantum communication, and the potential development of quantum computers.

By unraveling the mysteries of Schrödinger's Cat and exploring the depths of quantum superposition and entanglement, we enter a realm of profound questions and limitless possibilities. These concepts challenge our perception of reality and pave the way for groundbreaking advancements in computing, communication, and cryptography through the utilization of quantum technologies.

As we journey further into the fascinating world of quantum physics, we will continue to uncover extraordinary phenomena, including quantum measurement, quantum teleportation, and the philosophical implications that arise from this enigmatic field.

Embracing Uncertainty: The Heisenberg Uncertainty Principle

In our exploration of quantum mechanics, we encounter a fundamental principle that revolutionized our understanding of the microscopic world: the Heisenberg Uncertainty Principle. Proposed by Werner Heisenberg in 1927, this principle unveils the inherent limits in our ability to simultaneously measure certain pairs of physical properties with precision.

The Heisenberg Uncertainty Principle states that there is a fundamental limit to the precision with which we can know both the position and momentum of a particle. In simple terms, the more precisely we try to determine the position of a particle, the less precisely we can know its momentum, and vice versa. This principle is not a reflection of experimental limitations or shortcomings in our measurement devices; rather, it is an inherent feature of the quantum nature of reality.

To comprehend the significance of the Heisenberg Uncertainty Principle, we need to explore its implications in various aspects of quantum mechanics. One of its key consequences is that it introduces an element of randomness and unpredictability into the behavior of quantum systems. Unlike classical physics, where the properties of particles can be precisely determined, quantum particles exhibit inherent uncertainty in their behavior.

This uncertainty arises from the wave-like nature of quantum particles. As we discussed earlier, particles can exhibit both wave and particle characteristics, and this duality becomes particularly pronounced when considering the uncertainty principle. The position of a particle is associated with its wave function, which describes the probability distribution of finding the particle in different locations. The more spread out the wave function, the less precisely we can determine the particle's position.

Similarly, the momentum of a particle is linked to the wavelength of its associated wave function. The more precisely we try to determine the momentum of a particle, the more spread out its wave function becomes in terms of spatial extent. This leads to a corresponding increase in uncertainty in the particle's position.

The Heisenberg Uncertainty Principle challenges our classical intuitions, where we are accustomed to thinking of particles as having definite positions and velocities. In the quantum realm, the precise values of these properties are inherently unknowable, and we can only determine probabilities of finding particles in certain states.

This principle has profound implications across various fields of physics. It places limits on our ability to simultaneously measure quantities such as position and momentum, energy and time, and angular momentum components. These limitations are not a result of technological constraints but are intrinsic to the fabric of the quantum world.

The Heisenberg Uncertainty Principle also has implications beyond the realm of measurement. It is intimately connected to the concept of wave-particle duality and the wave nature of quantum objects. It highlights the fundamental indeterminacy and unpredictability that underlie quantum systems.

Moreover, the uncertainty principle has practical applications. It forms the basis for technological advancements such as electron microscopy, where the diffraction of electrons off a sample provides information about its structure and composition. Additionally, it plays a crucial role in quantum cryptography, where the inherent uncertainty in measuring quantum states is leveraged to ensure secure communication.

The Heisenberg Uncertainty Principle invites us to embrace the inherent uncertainty and indeterminacy of the quantum world. It challenges our desire for absolute knowledge and invites us to navigate a realm of probability and potentiality. It reminds us that the very act of measurement perturbs the system being measured, and that our observations are intrinsically intertwined with the phenomena we seek to understand.

As we continue our exploration of quantum mechanics, we will delve deeper into the remarkable phenomena that emerge from the interplay of wave-particle duality, superposition, entanglement, and uncertainty. These concepts, although initially perplexing, form the foundation of a new understanding of the quantum world and have led to groundbreaking technological advancements. By embracing the Heisenberg Uncertainty Principle, we open ourselves to the wonders and mysteries that quantum mechanics has to offer.

One fascinating aspect of the uncertainty principle is its connection to energy and time. The principle states that there is a fundamental limit to the precision with which we can simultaneously measure the energy of a quantum system and the time at which that measurement is made. This implies that the more accurately we try to determine the energy of a particle, the less precisely we can know the time at which the measurement occurs, and vice versa.

The uncertainty in energy and time arises from the intrinsic fluctuations and fluctuations in the quantum world. These fluctuations are rooted in the uncertainty principle and have profound implications for various phenomena. For instance, they contribute to the concept of virtual particles that pop in and out of existence in the vacuum of space. These virtual particles, although fleeting, have measurable effects and play a significant role in quantum field theory.

Furthermore, the uncertainty principle has implications for the stability and decay of particles. It introduces a natural limit to the lifetime of certain particles, as the uncertainty in their energy implies an inherent uncertainty in their lifespan. This principle underlies the concept of radioactive decay and sheds light on the dynamics of unstable particles.

The Heisenberg Uncertainty Principle also finds application in the realm of quantum entanglement. Entanglement is a phenomenon where two or more particles become intrinsically connected, such that the state of one particle is instantaneously correlated with the state of the other, regardless of the distance between them. The uncertainty principle plays a crucial role in quantifying the degree of entanglement between particles.

In addition to its theoretical implications, the uncertainty principle has practical consequences for technology. It poses challenges in the development of precise measurement devices and in the quest for absolute knowledge of quantum systems. However, it also inspires innovative solutions and novel technologies. For example, the development of quantum sensors and detectors takes into account the limitations imposed by the uncertainty principle and seeks to optimize measurement techniques within these bounds.

The Heisenberg Uncertainty Principle has sparked debates and discussions among physicists and philosophers alike. It raises fundamental questions about the nature of reality, the limits of human knowledge, and the interplay between observation and the observed. It challenges our classical intuitions and forces us to reevaluate our understanding of causality and determinism.

As we conclude our exploration of the Heisenberg Uncertainty Principle, we are reminded of the profound shift in our perspective that quantum mechanics brings. It unveils a world where certainty gives way to probability, where determinism gives way to indeterminacy, and where our understanding is shaped by the inherent limits imposed by the fabric of the quantum realm.

In the upcoming chapters, we will continue our journey into the depths of quantum mechanics. We will explore the phenomenon of quantum entanglement, the concept of quantum superposition, and the transformative potential of quantum computing. Together, these concepts will unravel the mysteries and unveil the remarkable possibilities that lie within the fascinating world of quantum physics.

Chapter 4: Exploring Quantum Theories

Quantum Field Theory: Unifying Forces and Particles

In our journey through the fascinating realm of quantum physics, we have delved into the enigmatic world of quantum mechanics, uncovering its fundamental principles and remarkable phenomena. As we progress, we now turn our attention to the profound framework of Quantum Field Theory (QFT), a powerful theoretical framework that unifies our understanding of forces and particles in the quantum realm.

Quantum Field Theory provides us with a captivating perspective on the nature of matter and its interactions. It encompasses the principles of quantum mechanics and incorporates the principles of special relativity, enabling us to describe the behavior of particles and their interactions in a unified manner.

At the heart of Quantum Field Theory lies the concept of a quantum field. In classical physics, we often think of particles as discrete objects moving through space and time. However, in the quantum realm, particles are understood as excitations or disturbances in their respective quantum fields. These fields permeate all of space-time and are present everywhere, interacting and exchanging energy and information.

Each fundamental particle in the universe is associated with a specific quantum field. For example, the electron is associated with the electron field, while the photon is associated with the electromagnetic field. These fields are not static but are dynamic and constantly fluctuating, giving rise to the particles and their interactions that we observe.

Quantum Field Theory describes these fields and their interactions using mathematical equations and principles. It allows us to calculate probabilities of particle interactions and predict the outcomes of experiments. Through the use of Feynman diagrams, which represent the various possible interactions between particles, we can visualize and calculate the probabilities of different particle processes.

One of the remarkable achievements of Quantum Field Theory is the unification of the electromagnetic and weak nuclear forces into the Electroweak Theory. This theory describes the electromagnetic force responsible for interactions between charged particles and the weak nuclear force, which governs processes such as radioactive decay. The Electroweak Theory successfully unifies these forces and provides a deeper understanding of the underlying symmetries and interactions in the quantum world.

Another groundbreaking aspect of Quantum Field Theory is the incorporation of the strong nuclear force, which binds protons and neutrons together in atomic nuclei. This force is described by the theory of Quantum Chromodynamics (QCD). QCD introduces the concept of quarks, the elementary particles that make up protons and neutrons, and gluons, the carriers of the strong force. Through Quantum Field Theory, we gain insights into the dynamics of these particles and their interactions, leading to a deeper understanding of the structure of matter.

Furthermore, Quantum Field Theory provides a framework for addressing the nature of particles and their properties, including their masses and charges. Through the process of renormalization, which involves carefully accounting for and canceling out infinities that arise in calculations, we can obtain meaningful and finite results. This process allows us to extract physical information about particles and their interactions from the underlying quantum field theories.

The development and application of Quantum Field Theory have revolutionized our understanding of the fundamental forces and particles that shape the universe. From the electromagnetic and weak forces to the strong nuclear force, this framework provides a unified description of these interactions, offering profound insights into the nature of matter and the fabric of space-time.

In the subsequent chapters, we will continue our exploration of Quantum Field Theory, diving deeper into its mathematical formalism and its applications in understanding phenomena such as particle decay, scattering processes, and the fundamental symmetries of the universe. We will also explore the concept of the Higgs field and its role in particle mass generation, shedding light on one of the most intriguing aspects of the quantum world.

As we embark on this exciting journey into the depths of Quantum Field Theory, we are poised to unravel the mysteries of the quantum realm and gain a profound understanding of the fundamental building blocks of the universe. Quantum Field Theory not only provides us with a powerful tool for calculation and prediction but also offers a profound philosophical perspective on the nature of reality.

One of the key concepts in Quantum Field Theory is the idea of particle creation and annihilation. According to the theory, particles are constantly being created and annihilated in the quantum fields, resulting in a dynamic and ever-changing picture of the quantum world. These processes, governed by the principles of quantum mechanics, give rise to the intricate web of interactions that we observe in nature.

To describe the behavior of particles in Quantum Field Theory, we use mathematical objects known as operators. These operators act on the quantum fields, transforming them and allowing us to calculate the probabilities of different particle processes. The interaction between particles is represented by terms in the mathematical equations that describe the theory, and through careful calculations, we can determine the likelihood of different particle interactions.

Quantum Field Theory also introduces the concept of symmetries, which play a fundamental role in understanding the laws of nature. Symmetries describe the invariance of physical systems under certain transformations, such as rotations, translations, and gauge transformations. By studying the symmetries of quantum fields and their interactions, we gain deep insights into the underlying structure of the universe.

One of the remarkable achievements of Quantum Field Theory is the successful unification of the electromagnetic and weak forces into the Electroweak Theory, as mentioned earlier. This theory is based on the principle of gauge symmetry and provides a unified description of electromagnetic and weak interactions. It explains phenomena such as the behavior of particles in particle accelerators and the process of particle decay.

Another intriguing aspect of Quantum Field Theory is its connection to the concept of particles and antiparticles. According to the theory, particles and antiparticles are two manifestations of the same underlying quantum field. Particle-antiparticle pairs can be created and annihilated through interactions with other particles or fields, resulting in fascinating phenomena such as particle decay and pair production.

The mathematical formalism of Quantum Field Theory is complex and requires advanced techniques from mathematical physics. It involves the use of integrals, differential equations, and sophisticated mathematical concepts such as functional analysis and group theory. These tools enable us to manipulate the quantum fields and calculate the probabilities of particle interactions with remarkable precision.

Through experiments and observations, scientists have confirmed the predictions of Quantum Field Theory and its ability to accurately describe the behavior of particles and their interactions. From the discovery of the Higgs boson at the Large Hadron Collider to the precise calculations of the anomalous magnetic moment of the electron, Quantum Field Theory has consistently proven its efficacy in explaining the fundamental phenomena of the quantum world.

In summary, Quantum Field Theory stands as a profound and elegant framework that unifies our understanding of forces and particles in the quantum realm. It provides us with a powerful mathematical language to describe the behavior of particles, their interactions, and the underlying symmetries of nature. With each new discovery and advancement in our knowledge, we gain a deeper appreciation for the intricate and wondrous nature of the quantum world.

In the upcoming chapters, we will delve further into the intricacies of Quantum Field Theory, exploring topics such as quantum electrodynamics, quantum chromodynamics, and the Standard

Model of particle physics. We will uncover the secrets of particle interactions, delve into the mysteries of quantum fluctuations, and examine the profound implications of Quantum Field Theory for our understanding of the universe.

As we continue our exploration of quantum theories, let us embrace the beauty and complexity of the quantum world, where particles and fields dance in harmony, and the laws of physics reveal their true quantum nature.

Seeking Unity at the Cosmic Scale: Quantum Gravity

In our quest to comprehend the fundamental nature of the universe, we encounter a formidable challenge when we try to merge the principles of quantum mechanics with the theory of general relativity. While quantum mechanics successfully describes the behavior of particles on microscopic scales and general relativity provides a remarkable framework for understanding gravity on cosmic scales, the two theories appear incompatible when we attempt to unify them.

This pursuit of uniting quantum mechanics and general relativity has led us to the realm of Quantum Gravity. Quantum Gravity seeks to reconcile the seemingly contradictory principles of quantum mechanics and general relativity by formulating a consistent theory that describes the nature of gravity at the quantum level. It aims to provide a unified framework that encompasses the behavior of particles and the curvature of space-time.

One of the major challenges in the development of Quantum Gravity is the quantization of gravity itself. In quantum mechanics, particles are described by wave functions and treated as discrete entities, while in general relativity, gravity is understood as the curvature of space-time caused by the presence of matter and energy. Quantizing gravity involves treating the gravitational field as a quantum field and quantizing its interactions with matter and other fields.

Various approaches have been proposed in the quest for a theory of Quantum Gravity. One prominent approach is String Theory, which suggests that elementary particles are not point-like entities but rather tiny, vibrating strings. These strings exist in higher-dimensional spaces, and their vibrations give rise to different particles and their interactions. String Theory offers the potential to reconcile quantum mechanics and gravity within a single framework.

Another approach is Loop Quantum Gravity, which takes a different path towards quantizing gravity. It proposes that space-time is made up of discrete, quantized entities known as loops or spin networks. These loops represent the fundamental building blocks of space-time, and their interactions give rise to the geometry and curvature associated with gravity. Loop Quantum Gravity provides insights into the nature of space-time on extremely small scales and offers a different perspective on the quantum nature of gravity.

Quantum Gravity also intersects with other areas of research, such as black hole physics and the study of the early universe. The behavior of black holes, with their extreme gravitational forces and singularities, poses intriguing questions about the interplay between quantum mechanics and gravity. Quantum Gravity aims to shed light on the quantum properties of black holes and the nature of space-time near their event horizons.

Furthermore, Quantum Gravity plays a crucial role in our understanding of the early universe. The birth of the universe itself, as described by the Big Bang theory, requires a quantum description of gravity to fully grasp the dynamics of the expanding cosmos. By investigating the quantum nature of the early universe, Quantum Gravity offers insights into the origin of the universe, the formation of structures, and the fundamental forces that shape our cosmos.

Despite significant progress, Quantum Gravity remains an active field of research with many open questions and ongoing debates. Scientists continue to explore different theoretical frameworks, develop mathematical tools, and conduct experiments to probe the boundaries of our understanding. The unification of quantum mechanics and general relativity at the quantum level represents a grand intellectual endeavor that could revolutionize our understanding of the cosmos.

In the upcoming chapters, we will dive deeper into the intricacies of Quantum Gravity, exploring the mathematical formalisms, the implications for our understanding of space-time, and the ongoing efforts to test and validate different approaches. We will delve into the fascinating concepts of black hole thermodynamics, quantum cosmology, and the quest for a complete theory of Quantum Gravity.

As we venture further into the realm of Quantum Gravity, let us embrace the inherent challenges and mysteries that lie ahead. The quest for a unified theory of quantum mechanics and gravity represents a profound intellectual journey that pushes the boundaries of human knowledge and uncovers the deep connections between the microcosm and the macrocosm. It is a quest that requires us to think beyond the boundaries of conventional wisdom and embrace the profound implications of a quantum universe.

One of the fundamental aspects of Quantum Gravity is the concept of space-time quantization. According to this idea, space-time itself is not continuous but is composed of discrete, indivisible units. Just as quantum mechanics introduced the notion of quantized energy levels, Quantum Gravity posits the existence of quantized units of space-time. These tiny units, often referred to as "quantum foam," form the fabric of the universe, vibrating and interacting in a complex dance that gives rise to the phenomena we observe.

The implications of space-time quantization are far-reaching. They suggest that at the most fundamental level, reality is granular, with a discrete structure that underlies the smoothness and continuity we perceive. This concept challenges our classical understanding of space and time and opens up new possibilities for exploring the nature of the cosmos.

In the pursuit of Quantum Gravity, physicists have encountered intriguing phenomena, such as the holographic principle and the information paradox. The holographic principle suggests that the information contained within a region of space can be encoded on its boundary, challenging our intuitions about the relationship between volume and information storage. The information paradox, on the other hand, arises when considering the fate of information that falls into a black hole, leading to questions about the conservation of information and the nature of black hole evaporation.

The study of Quantum Gravity also sheds light on the concept of quantum entanglement and its role in the fabric of space-time. Entanglement, a phenomenon where the properties of two particles become intertwined regardless of their separation, is a cornerstone of quantum mechanics. In the context of Quantum Gravity, entanglement is thought to play a crucial role in the underlying structure of space-time, providing a deeper understanding of the interconnectedness of the universe.

Moreover, Quantum Gravity offers a unique perspective on the nature of time. In classical physics, time flows uniformly and independently of other physical processes. However, in the quantum realm, time becomes entangled with other variables, leading to the concept of "quantum time." Quantum Gravity investigates the quantum nature of time and explores whether time itself emerges from more fundamental quantum processes.

The exploration of Quantum Gravity is not limited to theoretical considerations. Experimental investigations play a crucial role in validating and refining our understanding of the quantum nature of gravity. From probing the properties of black holes and studying the remnants of cosmic events to designing sensitive detectors and conducting high-energy experiments, scientists are pushing the boundaries of technology and knowledge to unravel the mysteries of Quantum Gravity.

In conclusion, the quest for Quantum Gravity represents a fascinating and profound journey into the heart of the universe. It requires us to bridge the gap between quantum mechanics and general relativity, delving into the nature of space-time, gravity, and the fundamental building blocks of reality. With each step forward, we gain deeper insights into the interconnectedness and quantum fabric of the cosmos. As we continue our exploration, let us embrace the wonder and complexity of Quantum Gravity and strive to unravel the ultimate secrets of the universe.

String Theory: Harmonies in a Multidimensional Universe

In the quest to understand the fundamental nature of the universe, physicists have turned to a theory that promises to unify all the known forces of nature: String Theory. This remarkable framework goes beyond the traditional notion of particles as point-like objects and introduces the concept of tiny, vibrating strings as the fundamental building blocks of reality.

At its core, String Theory suggests that the fundamental constituents of the universe are not zero-dimensional particles but rather one-dimensional strings. These strings can vibrate at different frequencies, giving rise to the diverse particles and forces we observe in the universe. Just as the harmonics of a vibrating string produce different musical notes, the vibrations of these fundamental strings generate the particles with distinct masses and properties.

One of the key features of String Theory is its requirement for additional dimensions beyond the familiar three spatial dimensions and one time dimension. In fact, String Theory proposes that the universe consists of not just the four dimensions we perceive but rather a total of ten or even eleven dimensions. These extra dimensions, often referred to as "compactified" dimensions, are curled up and hidden from our everyday experience at microscopic scales.

The introduction of these extra dimensions in String Theory allows for a richer and more comprehensive description of the universe. It provides a framework that can potentially reconcile the principles of quantum mechanics with the geometry of space-time. Moreover, the intricate interplay between the vibrational modes of the strings and the geometry of the extra dimensions gives rise to a vast landscape of possible physical configurations, offering a potential explanation for the diversity and complexity of our universe.

String Theory also offers an intriguing perspective on the nature of gravity. In this framework, gravity emerges naturally as the manifestation of the interactions between strings. Unlike other quantum field theories, where gravity is treated as a separate force, String Theory incorporates gravity within its fundamental framework, providing a unified description of all the forces of nature.

The elegance and mathematical beauty of String Theory have captivated physicists for decades. Yet, despite its promise, the theory still faces significant challenges. One of the major hurdles is the issue of experimental verification. Due to the extremely high energies required to directly detect strings or observe the effects of the extra dimensions, experimental confirmation of String Theory has remained elusive.

Nevertheless, String Theory has had a profound impact on our understanding of physics and has led to important advancements in related fields. It has provided new insights into black hole physics, the nature of quantum gravity, and the behavior of matter under extreme conditions. Furthermore, the mathematical techniques and concepts developed in String Theory have found applications in other areas of theoretical physics and mathematics, contributing to our broader understanding of the universe.

In recent years, String Theory has evolved into various branches and approaches, such as M-theory, brane-world scenarios, and holography. These developments have expanded our understanding of the theory and its connections to other areas of physics, including cosmology and particle physics.

As we continue to explore the intricate harmonies and symmetries within String Theory, we strive to uncover the deeper truths about the nature of the universe. The search for experimental evidence, the refinement of mathematical techniques, and the pursuit of deeper insights into the theory's fundamental principles are ongoing endeavors that will shape the future of theoretical physics.

In conclusion, String Theory stands as a powerful and elegant framework that seeks to unravel the mysteries of the universe. With its concept of vibrating strings and hidden dimensions, it offers a new perspective on the nature of particles, forces, and space-time. While challenges remain, the pursuit of String Theory continues to inspire and challenge scientists, pushing the boundaries of our understanding and opening new frontiers in the exploration of the quantum world.

Quantum Entanglement: Spooky Connections and Teleportation

In the fascinating realm of quantum mechanics, a phenomenon known as quantum entanglement has captured the attention of scientists and challenged our understanding of reality. Quantum entanglement refers to the intricate and mysterious correlations that can exist between particles, even when they are separated by vast distances. This phenomenon has been described by Albert Einstein as "spooky action at a distance" due to its seemingly non-local nature.

At the heart of quantum entanglement lies the principle of superposition, which allows particles to exist in multiple states simultaneously. When two particles become entangled, their quantum states become inseparably linked, regardless of the distance between them. This means that measuring the state of one particle instantly determines the state of the other, regardless of the spatial separation.

The implications of quantum entanglement are profound. It challenges our intuitive notions of cause and effect and suggests the existence of a hidden interconnectedness at the fundamental level of reality. Furthermore, quantum entanglement has been experimentally verified and has important applications in various fields, including quantum computing, cryptography, and teleportation.

One of the most remarkable applications of quantum entanglement is quantum teleportation. Teleportation, as depicted in science fiction, involves the instantaneous transportation of matter from one location to another. While the teleportation of physical objects is currently beyond our technological capabilities, quantum teleportation allows for the transfer of information between entangled particles.

The process of quantum teleportation begins with two entangled particles, often referred to as the "entangled pair." One of these particles, known as the sender or Alice, holds the information that needs to be teleported, while the other particle, called the receiver or

Bob, is located at the destination where the information is to be transferred.

To initiate the teleportation, Alice performs a measurement on her particle, which collapses its quantum state and transfers the information to Bob's particle instantaneously. However, due to the nature of quantum mechanics, this measurement is non-destructive, meaning that the original information remains intact in Alice's particle. Bob then performs a series of operations based on the measurement results from Alice, effectively reconstructing the quantum state of the teleported information.

Quantum teleportation is a remarkable demonstration of the power of entanglement and the non-local correlations it enables. It showcases the potential for transmitting information in a way that is secure, as any attempt to intercept or tamper with the entangled particles would disrupt the delicate entanglement and be detectable.

The phenomenon of quantum entanglement has led to deep theoretical investigations and has sparked debates about the nature of reality itself. The concept challenges classical notions of locality and separability, raising questions about the fundamental structure of the universe. It has also stimulated the development of new mathematical frameworks, such as quantum information theory, to describe and quantify the properties of entangled systems.

Moreover, quantum entanglement has connections to other areas of physics, such as black hole physics and the study of condensed matter systems. It has also been explored in the context of fundamental principles like the holographic principle and the AdS/CFT correspondence, which provide insights into the nature of gravity and the relationships between quantum field theories and gravitational theories.

The exploration of quantum entanglement continues to captivate researchers and drive scientific progress. Experimental studies strive to push the limits of entanglement, testing the boundaries of quantum correlations and exploring the phenomenon in increasingly complex systems. Theoretical investigations aim to deepen our understanding of entanglement and its implications for the fabric of space-time and the nature of information itself.

In conclusion, quantum entanglement represents a fascinating and enigmatic aspect of quantum mechanics. Its ability to establish non-local correlations challenges our classical notions of reality and opens up new possibilities for information processing and communication. The phenomenon of quantum entanglement has been experimentally observed and its applications, such as quantum teleportation, have been demonstrated in the laboratory.

Quantum entanglement serves as a crucial foundation for emerging technologies like quantum computing and quantum communication. Quantum computers leverage the power of entanglement to perform computations that are beyond the capabilities of classical computers. The entangled qubits in a quantum computer can be manipulated to store and process information in a way that exploits the inherent parallelism and superposition of quantum states.

Furthermore, the field of quantum communication aims to utilize entanglement to achieve secure and unbreakable encryption protocols. By encoding information in entangled particles, it becomes impossible for eavesdroppers to intercept or tamper with the transmitted data without disrupting the delicate entanglement and alerting the communicating parties. This promises a new era of secure communication where information can be exchanged with absolute privacy.

The study of quantum entanglement also sheds light on the fundamental nature of reality. It challenges the classical concept of local realism, which assumes that physical properties exist independently of observation and that information cannot propagate faster than the speed of light. Entanglement violates these assumptions, suggesting the existence of a deep interconnection between quantum systems regardless of their spatial separation.

Moreover, quantum entanglement is closely linked to the concept of quantum non-locality, where the correlations between entangled particles extend beyond classical limits. This has led to the formulation of Bell's theorem and subsequent experimental tests that confirm the violation of Bell inequalities, providing strong evidence against local realistic theories.

Quantum entanglement has also sparked theoretical debates and investigations into its connection with other areas of physics, such as the study of black holes and the nature of gravity. The holographic principle, which relates a higher-dimensional gravitational theory to a lower-dimensional quantum field theory, suggests that entanglement plays a fundamental role in the description of space-time itself.

In recent years, there have been exciting developments in the field of quantum entanglement, including the exploration of multipartite entanglement involving more than two particles, the study of long-range entanglement in condensed matter systems, and the investigation of entanglement in high-energy physics experiments.

The quest to understand and harness the power of quantum entanglement is ongoing. Researchers continue to delve into the intricacies of entangled systems, seeking to unlock new insights and applications. As our knowledge and control over quantum phenomena advance, the potential for revolutionary technologies and a deeper understanding of the fundamental nature of the universe grows.

In conclusion, quantum entanglement is a cornerstone of quantum mechanics, defying classical intuitions and offering a profound perspective on the interconnectedness of the quantum world. It holds the key to groundbreaking technologies and continues to inspire scientific exploration, pushing the boundaries of our understanding and paving the way for a quantum future.

Chapter 5: Real-World Applications of Quantum Physics

Ultra-Precise Clocks: Timekeeping in the Quantum Realm

Time is a fundamental aspect of our existence, and the accurate measurement of time has been a quest for humanity throughout history. In the realm of quantum physics, the pursuit of precision has led to the development of ultra-precise clocks that push the boundaries of timekeeping accuracy.

Quantum clocks, based on the principles of quantum mechanics, offer an unprecedented level of accuracy that surpasses traditional clocks. These clocks utilize the unique properties of quantum systems to achieve extraordinary precision in timekeeping.

At the heart of quantum clocks are quantum oscillators, which are capable of existing in multiple states simultaneously. By harnessing the phenomenon of quantum superposition, scientists have developed clocks that can maintain a high degree of coherence and stability.

One such example is the development of atomic clocks, which rely on the behavior of atoms to measure time. In atomic clocks, atoms are carefully prepared and manipulated to ensure that they oscillate at a specific frequency. The oscillations of these atoms serve as the basis for measuring time intervals.

The exceptional accuracy of atomic clocks is made possible by the quantized nature of atomic energy levels. According to quantum mechanics, energy levels in atoms are discrete, and the transitions between these levels occur at specific frequencies. By precisely measuring the frequency of these transitions, scientists can determine the duration of a given time interval with remarkable accuracy.

To achieve even higher precision, scientists have employed techniques such as laser cooling and trapping of atoms. By cooling atoms to temperatures close to absolute zero, their motion is greatly reduced, minimizing the effects of external disturbances. This allows for more accurate time measurements and improved clock stability.

Quantum entanglement also plays a crucial role in enhancing the accuracy of quantum clocks. Entanglement enables the correlation of quantum states between particles, even when they are physically separated. By entangling multiple particles in a clock system, scientists can reduce the effects of noise and external influences, leading to more precise timekeeping.

The application of ultra-precise clocks extends beyond the measurement of time itself. These clocks have far-reaching implications in fields such as navigation, telecommunications, and scientific research.

In navigation systems like GPS, precise timekeeping is crucial for determining the position of objects on Earth's surface. Quantum clocks have revolutionized GPS technology by providing more accurate timing signals, resulting in improved positioning capabilities. This advancement has paved the way for applications such as autonomous vehicle navigation, precise synchronization of communication networks, and geodetic measurements.

The impact of quantum clocks extends to the field of fundamental physics as well. They are used to test theories such as general relativity and investigate the nature of fundamental constants. The unprecedented accuracy of quantum clocks allows scientists to probe the fabric of space-time and search for any deviations from established theories, providing valuable insights into the fundamental laws of the universe.

In the realm of metrology, the science of measurement, quantum clocks serve as essential references for calibrating and verifying other measuring instruments. Their precision is crucial in industries where accurate timekeeping is vital, including telecommunications, financial systems, and scientific experiments requiring precise synchronization.

Looking ahead, ongoing research in quantum clock technology continues to push the boundaries of precision timekeeping. Developments such as optical clocks, which operate at even higher frequencies using laser light, hold the promise of even greater accuracy and stability. Optical clocks have the potential to redefine our understanding of time and find applications in fields such as satellite navigation, high-precision tests of fundamental physics, and the synchronization of global communication networks.

In conclusion, quantum physics has revolutionized the field of timekeeping with the development of ultra-precise clocks. By harnessing the principles of quantum mechanics, these clocks offer unprecedented accuracy and stability. They have found applications in navigation, telecommunications, scientific research, and various other industries where precise timekeeping is essential.

One of the notable applications of ultra-precise quantum clocks is in the field of satellite-based navigation systems, such as the Global Positioning System (GPS). GPS relies on accurate time synchronization to determine the position of receivers on Earth's surface. By employing quantum clocks, which provide incredibly precise timing signals, the accuracy of GPS positioning has been significantly enhanced. This advancement has led to improvements in navigation for various applications, including aviation, maritime, and land-based transportation systems. It enables precise tracking, efficient routing, and reliable location-based services, benefiting industries and individuals worldwide.

Telecommunications is another domain where the impact of quantum clocks is profound. Communication networks heavily rely on precise timing synchronization to ensure reliable data transmission and coordination among interconnected devices. Quantum clocks offer highly accurate time references that enable precise synchronization of communication networks, enhancing the efficiency and reliability of data transfer. From mobile networks to internet infrastructure, quantum clocks play a crucial role in maintaining seamless connectivity and enabling real-time communication.

In scientific research, ultra-precise clocks are invaluable tools for investigating fundamental phenomena and pushing the boundaries of our understanding. Quantum clocks contribute to experiments that test the foundations of physics, such as the search for violations of the laws of physics or the study of the nature of gravity. The remarkable accuracy of these clocks allows scientists to perform high-precision measurements, enabling them to detect subtle effects and potential deviations from established theories. Quantum clocks also find applications in areas like cosmology, where they assist in studying the evolution of the universe, cosmic microwave background radiation, and the measurement of cosmic distances.

Moreover, the advances in quantum clock technology have implications for various industries where precise timing is crucial. Financial systems rely on accurate timekeeping for trading, transaction processing, and synchronization of global markets. Quantum clocks provide the precision required to ensure fairness, efficiency, and security in financial operations.

Furthermore, ultra-precise clocks are valuable in metrology, the science of measurement. They serve as primary references for calibrating and validating other measuring instruments. The exceptional accuracy and stability of quantum clocks enable highly precise measurements in fields such as metrology laboratories, manufacturing, and quality control. This, in turn, contributes to advancements in industries such as aerospace, engineering, and scientific research, where precise measurements are essential for ensuring safety, reliability, and innovation.

Looking forward, the ongoing advancements in quantum clock technology hold great promise for future applications. Researchers are actively exploring novel approaches, such as optical lattice clocks, quantum sensors, and advanced quantum algorithms, to further enhance the accuracy, stability, and portability of quantum clocks. These developments have the potential to revolutionize a wide range of fields, including space exploration, fundamental physics research, quantum computing, and even the redefinition of international standards for timekeeping.

In conclusion, ultra-precise quantum clocks, born out of the principles of quantum physics, have far-reaching applications in diverse fields. Their unrivaled accuracy and stability have revolutionized satellite navigation, telecommunications, scientific research, finance, metrology, and numerous other industries. As we continue to explore and advance quantum clock technology, we are poised to unlock new realms of precision and usher in an era of even more remarkable applications in the future.

Quantum Key Distribution: Securing Communication

In today's interconnected world, ensuring the security of communication has become of paramount importance. The exchange of sensitive information, whether it's financial transactions, personal data, or classified messages, demands robust encryption methods that can protect against unauthorized access and maintain the privacy of the involved parties. Quantum key distribution (QKD) stands at the forefront as an innovative solution that harnesses the principles of quantum physics to establish secure communication channels.

Traditional encryption techniques rely on complex mathematical algorithms, which, in theory, can be deciphered given sufficient computational power and time. However, QKD takes advantage of the fundamental properties of quantum mechanics, such as the Heisenberg uncertainty principle and quantum entanglement, to offer a level of security that is theoretically unbreakable.

At the heart of QKD lies the concept of quantum entanglement. Quantum entanglement enables the creation of correlations between two or more particles, such as photons, in a way that the state of one particle is instantaneously affected by the state of the other, regardless of the distance separating them. This remarkable phenomenon forms the basis for establishing a shared secret key between a sender, often referred to as Alice, and a receiver, known as Bob, which can be used to secure their communication.

The QKD protocol initiates with Alice preparing a series of entangled photon pairs and sending one photon from each pair to Bob over a dedicated quantum channel. This channel can be realized using various technologies, including fiber optic cables or even free space, depending on the implementation. Upon receiving the photons, Bob performs measurements on them using randomly chosen bases. Simultaneously, Alice, who is aware of the bases used, transmits the information about these bases to Bob through a classical communication channel.

The security of QKD relies on the properties of quantum measurements. Any attempt to eavesdrop on the quantum channel and gain information about the shared key would unavoidably disturb the delicate quantum state of the photons, thereby introducing detectable errors in Bob's measurements. By employing a process known as quantum error detection, Alice and Bob can identify the presence of an eavesdropper and discard any compromised key, ensuring the security of their communication.

Once the QKD process is successfully executed, Alice and Bob possess a shared secret key known exclusively to them. This key can subsequently be used in conjunction with conventional encryption algorithms, such as the Advanced Encryption Standard (AES), to encrypt and decrypt their messages. Notably, since the security of the key distribution relies on the fundamental laws of quantum mechanics, any attempts to intercept or tamper with the key during transmission would be immediately detected, guaranteeing the integrity and confidentiality of the communication.

One of the most remarkable aspects of QKD is its unconditional security. Unlike traditional encryption methods, which may become vulnerable to advancements in computing power or algorithmic breakthroughs, QKD offers provable security based on the fundamental principles of quantum mechanics. The security of the key distribution hinges upon the inherent impossibility of measuring certain properties of quantum particles without disturbing their states, rendering it virtually impossible for an eavesdropper to surreptitiously obtain the key.

The potential applications of QKD span a wide range of domains where secure communication is paramount. For instance, QKD can be deployed in government and military communications to safeguard classified information from unauthorized interception. It can also enhance the security of financial transactions, ensuring that sensitive data such as credit card information or online banking credentials remains protected during transmission. Moreover, QKD holds the potential to revolutionize healthcare systems by safeguarding patients' confidential medical records and enabling secure telemedicine services.

Despite significant advancements, several challenges remain to be addressed before QKD can achieve widespread adoption. One primary challenge is the limited range of quantum channels, which typically spans a few hundred kilometers. Efforts are currently underway to overcome this limitation by developing quantum repeaters and satellite-based systems that can extend the reach of QKD to intercontinental distances. These technological advancements would enable secure communication on a global scale, revolutionizing industries and facilitating secure interactions across borders.

Another challenge facing QKD adoption is the practical implementation and integration of the technology into existing communication infrastructures. While significant progress has been made in developing QKD systems, further research and development are needed to make them more compact, cost-effective, and compatible with conventional communication protocols. The seamless integration of QKD into existing networks will be crucial for its widespread deployment and usability in real-world scenarios.

Moreover, ensuring the reliability and stability of QKD systems is of utmost importance. Environmental factors, such as temperature fluctuations and noise, can introduce errors and reduce the efficiency of quantum communication. Therefore, ongoing research aims to develop robust and resilient QKD systems that can operate under various environmental conditions, ensuring consistent and secure communication even in challenging situations.

The applications of QKD extend beyond secure communication between two parties. The technology also enables secure multi-party communication, where multiple entities can establish shared secret keys, allowing for secure group communication and collaboration. This capability opens up new possibilities in areas such as secure video conferencing, secure cloud computing, and secure data sharing among multiple stakeholders.

Furthermore, QKD has the potential to play a significant role in the future development of quantum networks. Quantum networks, also known as quantum internet, aim to interconnect quantum devices and enable the transmission of quantum information over long distances. QKD serves as a critical component of quantum networks by providing the necessary secure key exchange between network nodes. With the development of quantum repeaters and advanced network protocols, quantum networks could revolutionize fields such as quantum computing, quantum sensing, and secure quantum cloud computing.

As with any emerging technology, there are also societal and ethical considerations associated with the widespread adoption of QKD. The use of secure communication technologies raises questions about privacy, surveillance, and the balance between security and individual freedoms. It is essential to address these concerns and establish transparent policies and regulations that safeguard both the security of communication and the privacy rights of individuals.

In conclusion, quantum key distribution (QKD) offers a groundbreaking approach to secure communication by leveraging the principles of quantum mechanics. Through the utilization of quantum entanglement and quantum measurements, QKD provides an unbreakable foundation for establishing shared secret keys between communicating parties. The practical applications of QKD span various domains, including government communications, financial transactions, healthcare systems, and beyond. Challenges, such as range limitations and integration into existing infrastructures, continue to be addressed through ongoing research and development efforts. As QKD progresses, it has the potential to revolutionize secure communication and pave the way for the future development of quantum networks and quantum-based technologies.

Quantum Computing: Unleashing the Power of Quantum Bits

Quantum computing stands at the forefront of scientific and technological advancements, heralding a new era of computational power and capabilities. By harnessing the unique properties of quantum mechanics, quantum computers have the potential to revolutionize industries, accelerate scientific discoveries, and solve complex problems that are currently beyond the reach of classical computers. In this chapter, we delve deeper into the realm of quantum computing and explore its real-world applications.

Quantum computing operates on the principles of quantum bits, or qubits, which possess the extraordinary ability to exist in multiple states simultaneously. This phenomenon, known as superposition, allows quantum computers to process vast amounts of information in parallel, enabling exponential computational speed and power. Unlike classical computers that rely on classical bits represented as 0s and 1s, quantum computers leverage the richness of quantum states to perform computations.

One of the most remarkable algorithms in the realm of quantum computing is Shor's algorithm. This groundbreaking algorithm has the potential to efficiently factor large numbers, a task that is notoriously challenging for classical computers. Factoring large numbers plays a crucial role in modern cryptography, and the advent of quantum computers capable of factoring large numbers efficiently poses a significant threat to cryptographic systems. However, it also presents an opportunity for the development of quantum-resistant cryptographic algorithms that can withstand attacks from quantum computers.

Beyond cryptography, quantum computing holds immense promise in the field of optimization. Many real-world problems, such as supply chain management, portfolio optimization, and logistics planning, involve a multitude of variables and complex calculations. Classical computers struggle to find optimal solutions within a reasonable time frame due to the exponential growth of possibilities. Quantum computers, with their inherent parallelism, can explore multiple potential solutions simultaneously, offering the potential to revolutionize optimization processes and drive unprecedented efficiency gains in various industries.

Moreover, quantum computing has the potential to transform scientific simulations and modeling. Quantum simulators, specialized quantum computers designed to emulate and study quantum systems, provide insights into the behavior of materials, chemical reactions, and biological processes that are otherwise inaccessible to classical computers. This capability opens up new avenues for drug discovery, materials design, and fundamental research in quantum phenomena.

Despite the immense potential, quantum computing faces formidable challenges on the path to widespread adoption and practicality. One of the primary challenges is qubit stability and coherence. Qubits are highly sensitive to noise and environmental disturbances, which can introduce errors and compromise the integrity of quantum computations. Researchers are actively investigating methods to mitigate these challenges, including error-correction techniques, advancements in qubit design and fabrication, and the development of robust control mechanisms to enhance the stability and coherence of qubits.

Another crucial challenge lies in scaling up the number of qubits. Presently, quantum computers with a few dozen qubits exist, but to tackle complex real-world problems, millions or even billions of qubits will be required. Achieving such scalability while maintaining qubit quality and coherence poses significant engineering and technological hurdles. Various approaches, such as improving qubit architectures, exploring alternative qubit implementations, and optimizing qubit connectivity, are being pursued to address this challenge.

Furthermore, the programming and control of quantum computers present unique complexities. Quantum algorithms and circuits differ fundamentally from classical programming paradigms, necessitating a deep understanding of quantum mechanics and quantum information theory. Developing user-friendly programming languages, software frameworks, and intuitive interfaces that abstract the intricacies of quantum computing is crucial for accelerating the adoption and utilization of quantum computers by a broader range of researchers and developers.

Despite these challenges, remarkable progress has been made in the development of quantum computing technologies. Multiple hardware platforms, including superconducting qubits, trapped ions, and topological qubits, are continuously advancing, each with its own advantages and challenges. Furthermore, quantum software frameworks, such as Qiskit and Microsoft's Q#, provide accessible platforms for researchers and developers to explore and experiment with quantum algorithms and applications. These frameworks offer a range of tools and libraries for designing and simulating quantum circuits, implementing quantum algorithms, and interfacing with quantum hardware.

As the field of quantum computing continues to evolve, collaborations between academia, industry, and research institutions are becoming increasingly vital. These partnerships foster knowledge exchange, drive innovation, and accelerate the development of quantum technologies. Quantum computing companies and startups are emerging, aiming to build scalable quantum systems, develop quantum algorithms, and provide quantum-based solutions for various industries.

The impact of quantum computing extends beyond the realms of science and technology. Its potential applications are far-reaching and encompass diverse fields such as finance, healthcare, materials science, artificial intelligence, and more. Quantum machine learning algorithms can revolutionize data analysis, enabling more accurate predictions and insights. Quantum simulations can aid in drug discovery and molecular design, accelerating the development of new medications and materials. Furthermore, quantum-inspired optimization algorithms can optimize complex systems and improve decision-making processes.

Ethical considerations surrounding quantum computing are also emerging as the technology progresses. Quantum computing's immense computational power raises concerns about the security and privacy of sensitive information. Cryptographic systems, data encryption methods, and cybersecurity protocols need to adapt to the potential threat posed by quantum computers. Ethical frameworks and regulations must be established to ensure responsible and secure use of quantum computing technologies.

Looking ahead, continued research and development efforts are essential to overcome the remaining challenges and unlock the full potential of quantum computing. Advancements in qubit stability, coherence, and scalability are crucial for building more robust and powerful quantum systems. The refinement of error-correction techniques and the development of fault-tolerant quantum computing architectures will pave the way for reliable and practical quantum computers.

Furthermore, interdisciplinary collaborations between quantum physicists, computer scientists, mathematicians, and engineers will be instrumental in expanding our understanding of quantum phenomena, developing novel quantum algorithms, and translating quantum research into practical applications. Educational programs and initiatives should be established to train the next generation of quantum scientists and engineers, fostering a skilled workforce capable of harnessing the power of quantum computing.

In conclusion, the fifth chapter of this book has explored the real-world applications of quantum physics, with a particular focus on quantum computing. Quantum computing offers unprecedented computational power and capabilities, revolutionizing fields such as optimization, cryptography, simulation, and machine learning. While challenges remain, significant progress has been made, and the prospects for quantum technologies are incredibly promising. By embracing quantum physics and its applications, we embark on a journey of discovery and innovation, reshaping the landscape of science, technology, and our understanding of the universe.

Chapter 6: Reflecting on the Quantum Journey

Key Insights and Takeaways

As we reflect on the fascinating journey we have embarked on through the realms of quantum physics, it is important to distill the key insights and takeaways that have emerged from our exploration. Throughout this book, we have delved into the fundamental principles, mind-bending concepts, and groundbreaking applications of quantum physics. Let us now consolidate our understanding and draw meaningful conclusions from our quantum odyssey.

One of the fundamental insights we have gained is the dual nature of particles, as illuminated by the wave-particle duality. The realization that particles can exhibit both wave-like and particle-like behaviors challenges our classical intuitions and forms the foundation of quantum mechanics. This insight has profound implications for our understanding of the microscopic world and the fundamental building blocks of the universe.

Another significant takeaway is the role of uncertainty in quantum physics, exemplified by the Heisenberg uncertainty principle. We have learned that there are inherent limitations to our ability to precisely measure certain pairs of physical properties of particles. This principle not only highlights the probabilistic nature of quantum phenomena but also underscores the intrinsic limits of our knowledge and the importance of statistical interpretations in quantum theory.

The concept of quantum entanglement has captivated our imagination and revealed the profound interconnectedness of quantum systems. The non-local correlations observed between entangled particles defy classical notions of locality and have spurred advancements in quantum communication and cryptography. The phenomenon of entanglement has provided a fertile ground for exploring the intriguing concept of quantum teleportation and has the potential to revolutionize secure communication protocols.

Moreover, we have explored the power and potential of quantum computing. Quantum bits, or qubits, with their ability to exist in superposition and entanglement, offer a radically different approach to computation. Quantum algorithms, such as Shor's algorithm for factoring large numbers and Grover's algorithm for database search, hold the promise of solving complex problems exponentially faster than classical computers. The development of fault-tolerant quantum architectures and error-correction techniques is crucial for realizing the full potential of quantum computing.

In our quantum journey, we have also encountered the profound implications of quantum physics on our understanding of time, space, and the fabric of the universe. The exploration of quantum gravity and string theory has offered glimpses into the unification of fundamental forces and the nature of spacetime at the cosmic scale. These theoretical frameworks strive to reconcile quantum mechanics with general relativity, and although many challenges remain, they represent bold steps toward a unified theory of physics.

The real-world applications of quantum physics have immense transformative potential across various domains. From ultra-precise clocks and quantum sensors that enable unprecedented measurements to quantum key distribution systems that provide secure communication channels, quantum technologies are poised to revolutionize industries and drive technological advancements. The integration of quantum physics with fields such as materials science, medicine, finance, and artificial intelligence holds the promise of addressing complex problems and unlocking new avenues of innovation.

As we reflect on our quantum journey, it is important to acknowledge the interdisciplinary nature of quantum physics. Collaboration among physicists, mathematicians, computer scientists, and engineers is critical for advancing the frontiers of quantum research and translating it into practical applications. Educational programs and initiatives should be established to nurture and train the next generation of quantum scientists and engineers, ensuring a skilled workforce that can harness the power of quantum technologies.

In conclusion, our exploration of quantum physics has taken us on a captivating voyage through the mysteries and wonders of the quantum realm. We have gained deep insights into the fundamental nature of particles, the principles of quantum mechanics, and the transformative potential of quantum technologies. This journey has challenged our preconceptions, expanded our horizons, and highlighted the immense possibilities that lie ahead. By embracing the principles and applications of quantum physics, we can unlock new frontiers of knowledge, drive technological advancements, and shape the future of science and technology.

Throughout our quantum journey, we have witnessed the intricate dance between particles and waves, the delicate balance of uncertainty, and the mysterious phenomena of entanglement and superposition. These concepts have revolutionized our understanding of the microscopic world and have paved the way for remarkable technological innovations.

The insights gained from quantum physics have not only deepened our understanding of the fundamental nature of reality but have also paved the way for practical applications that were once thought to be purely science fiction. Ultra-precise atomic clocks, for instance, have emerged as vital tools for modern navigation systems, global positioning, and synchronization of communication networks. By harnessing the precise ticking of atoms, we can measure time with unprecedented accuracy, enabling advancements in fields ranging from satellite navigation to financial transactions.

Quantum key distribution has emerged as a revolutionary approach to secure communication. By exploiting the principles of quantum mechanics, such as the non-cloning theorem and the uncertainty principle, quantum cryptography provides a foolproof method of transmitting encrypted messages. This technology holds the potential to protect sensitive data from eavesdropping and ensure the privacy and security of information exchanged in the digital realm.

The advent of quantum computing has sparked excitement and anticipation in both the scientific and technological communities. With the power of quantum bits, or qubits, harnessing the principles of superposition and entanglement, quantum computers have the potential to solve complex problems exponentially faster than classical computers. This computational power opens up new avenues for optimization, simulation, and machine learning, with implications for fields as diverse as drug discovery, financial modeling, and artificial intelligence.

Looking beyond the horizon, the field of quantum gravity represents an ambitious endeavor to unify the theories of quantum mechanics and general relativity. The quest for a theory that encompasses both the microscopic and macroscopic scales has led physicists to explore the profound nature of space-time and the fabric of the universe. String theory, in particular, offers a framework that reconciles these fundamental theories and opens up the possibility of a multidimensional universe. While these theories are still in their infancy and pose significant challenges, they provide a glimpse into a deeper understanding of the cosmos.

The potential of quantum physics is not limited to the realms of science and technology alone. It has the power to transform our perception of reality and challenge our philosophical and metaphysical beliefs. The concept of quantum consciousness, for instance, explores the possibility that the fundamental principles of quantum mechanics may play a role in the processes of cognition and consciousness. This provocative idea blurs the boundaries between the physical and the mental, inviting us to question the nature of our existence.

In conclusion, our journey through the realms of quantum physics has revealed a captivating tapestry of fundamental principles, mind-bending phenomena, and transformative applications. From the dual nature of particles to the intricacies of quantum entanglement, from the potential of quantum computing to the quest for a unified theory of physics, the quantum world has captivated our imagination and pushed the boundaries of human knowledge. By embracing the insights and applications of quantum physics, we can continue to unravel the mysteries of the universe, revolutionize technology, and shape the future of scientific exploration. The quantum journey has just begun, and the possibilities are limitless.

Embracing the Mysteries: The Beauty of Quantum Physics

As we delve deeper into the world of quantum physics, we cannot help but be captivated by its inherent beauty and profound mysteries. Beyond its practical applications and scientific advancements, quantum physics offers us a glimpse into the elegance and complexity of the universe. In this final chapter, we will explore the awe-inspiring beauty of quantum physics and how it continues to inspire scientists and philosophers alike.

One of the most striking aspects of quantum physics is its ability to challenge our intuition and conventional understanding of reality. The principle of superposition, for instance, suggests that particles can exist in multiple states simultaneously, defying classical notions of definiteness and determinism. This concept invites us to question the very nature of reality and forces us to reconsider our fundamental assumptions about the world we inhabit.

Another intriguing aspect of quantum physics is the phenomenon of quantum entanglement. When two particles become entangled, their states become correlated in a way that defies classical explanations. This mysterious connection, instantaneously maintained regardless of distance, has been aptly described by Einstein as "spooky action at a distance." The beauty of entanglement lies in its profound implications for our understanding of interconnectedness and the intricate web of relationships that exist in the fabric of the universe.

Quantum physics also reveals the deep interplay between observation and the observed. The act of measurement itself can alter the behavior of particles, leading to the famous observer effect. This connection between the observer and the observed challenges the notion of an objective reality independent of our perception and highlights the intricate relationship between the observer, the measurement apparatus, and the quantum system under investigation.

The mathematical language of quantum physics further contributes to its beauty. The formalism of quantum mechanics, with its complex numbers, wave functions, and operators, provides a powerful tool to describe and predict the behavior of quantum systems. The elegance and symmetry of the mathematical framework evoke a sense of awe and wonder, drawing parallels to the harmony and beauty found in art and music.

Beyond its scientific and mathematical allure, quantum physics also offers profound philosophical insights. It confronts us with questions about the nature of existence, the limits of human knowledge, and the role of consciousness in shaping reality. Some interpretations of quantum mechanics, such as the Copenhagen interpretation or the many-worlds interpretation, delve into the philosophical implications of quantum phenomena, inviting us to ponder the nature of truth, determinism, and the fabric of the cosmos.

In the pursuit of understanding quantum physics, scientists and philosophers have embarked on a quest to unravel the deepest mysteries of the universe. This quest, driven by curiosity and a desire to comprehend the fundamental nature of reality, continues to inspire generations of thinkers and innovators. The beauty of quantum physics lies not only in its practical applications but also in its capacity to ignite our imagination and expand our intellectual horizons.

As we conclude our journey through the quantum realm, we are left with a profound appreciation for the beauty and intricacy of the quantum world. Its mysteries and paradoxes challenge us to explore new frontiers of knowledge and to embrace the unknown with open minds. Quantum physics invites us to marvel at the elegant dance of particles, the entangled web of connections, and the delicate balance of uncertainty. It is through the pursuit of understanding and the celebration of its beauty that we continue to unlock the secrets of the quantum realm and push the boundaries of human knowledge.

In this book, we have embarked on a voyage through the wonders of quantum physics, from its early beginnings to its far-reaching implications. We have explored the fundamental principles that underpin this extraordinary field, delved into its mind-boggling phenomena, and witnessed the transformative applications that have emerged from its depths. The journey has been awe-inspiring and enlightening, challenging our preconceived notions and expanding our understanding of the universe.

Throughout our exploration, we have encountered the wave-particle duality, a fundamental concept that shattered classical understanding and introduced us to the fascinating nature of quantum entities. We have witnessed the mysterious behavior of particles through the lens of quantum superposition, where they can exist in multiple states simultaneously, defying our intuitive understanding of reality.

The journey has also taken us into the realm of quantum entanglement, where particles become intricately connected in ways that transcend classical explanations. The phenomenon of entanglement has opened doors to exciting possibilities, such as quantum teleportation and secure communication through quantum key distribution. These breakthroughs have the potential to revolutionize various fields, from cryptography to information processing.

Moreover, we have explored the mathematical formalism of quantum mechanics, with its complex numbers, wave functions, and operators. This elegant language allows us to describe and predict the behavior of quantum systems with remarkable precision. It is through the mathematical framework of quantum mechanics that we gain insights into energy levels, quantum jumps, and the probabilistic nature of quantum phenomena.

As we journeyed deeper into the quantum world, we encountered the Heisenberg uncertainty principle, which sets fundamental limits on the precision with which certain pairs of physical properties, such as position and momentum, can be simultaneously known. This principle challenges our classical notion of determinism and underscores the inherent uncertainty that exists at the quantum level.

In our exploration, we also touched upon the unifying quest for a theory of everything, which brings together the principles of quantum mechanics and general relativity. The pursuit of quantum gravity and string theory aims to reconcile the microscopic and macroscopic realms, bridging the gap between the quantum and the cosmic scales.

Our journey through the quantum realm has been filled with astonishing discoveries and profound questions. It has led us to ponder the nature of reality, the role of consciousness in shaping our observations, and the limits of human knowledge. Quantum physics has not only transformed our understanding of the physical world but has also inspired philosophical and metaphysical debates, pushing the boundaries of human thought.

As we reflect on the quantum journey, we recognize the beauty that lies within its mysteries. The intricate dance of particles, the interconnectedness of quantum entanglement, and the delicate balance of uncertainty all contribute to the profound allure of quantum physics. Its elegance and complexity remind us of the inherent beauty that exists in the tapestry of the universe.

In conclusion, the exploration of quantum physics has taken us on a remarkable voyage of discovery. We have witnessed the astonishing phenomena that govern the subatomic realm, grappled with the profound implications of quantum theory, and marveled at the intricate beauty of the quantum world. Our journey through the chapters of this book has provided a glimpse into the remarkable insights and practical applications that have emerged from the study of quantum physics.

As we embrace the mysteries and beauty of quantum physics, we are reminded of the boundless potential that lies within the universe. It is through continued exploration, curiosity, and open-mindedness that we will unravel deeper layers of understanding and unlock new frontiers of knowledge. The quantum journey is ongoing, and it is our collective endeavor to push the boundaries of scientific inquiry and deepen our appreciation for the intricate fabric of the cosmos.

In the end, may this book serve as a guide, igniting curiosity and inspiring the next generation of quantum pioneers. May it encourage us to embrace the mysteries, appreciate the beauty, and embark on our own journey of discovery in the captivating realm of quantum physics.

Bonus 1

Quantum Tunneling: Journeying Through Barriers. Unveiling the Phenomenon of Quantum Tunneling

In the vast and mysterious realm of quantum physics, a fascinating phenomenon known as quantum tunneling emerges, defying the traditional boundaries of classical physics. Quantum tunneling allows particles to traverse energy barriers that would be classically impassable, offering a remarkable glimpse into the wave-like behavior of particles and the intricate fabric of our quantum universe. Let us embark on a journey to explore the concept of quantum tunneling and unravel its profound implications.

At the heart of quantum tunneling lies the wave-particle duality, a central theme throughout our exploration of quantum physics. This duality reveals that particles, such as electrons, can exist in multiple states simultaneously, exhibiting both particle-like and wave-like properties. It is this wave-like nature that allows particles to penetrate energy barriers, akin to a surfer riding a wave through a seemingly impenetrable obstacle.

Quantum tunneling challenges our classical intuitions, where we expect particles to follow well-defined trajectories dictated by classical physics. Instead, we encounter a realm governed by probabilities and uncertainty, where particles have the astonishing ability to "tunnel" through potential energy barriers, transcending the limitations imposed by classical physics. This remarkable phenomenon highlights the probabilistic nature of quantum systems, emphasizing the inherent uncertainty that lies at the heart of the quantum world.

The explanation for quantum tunneling lies in the wave nature of particles and the concept of wavefunction. According to quantum mechanics, particles are described by wavefunctions that represent the probability distribution of their positions. When a particle encounters an energy barrier, its wavefunction extends into the region beyond the barrier, allowing for a finite probability of finding the particle on the other side. This probability of tunneling decreases exponentially with increasing barrier height and width, reflecting the diminishing likelihood of overcoming larger and wider obstacles.

The implications of quantum tunneling reach far and wide, permeating various fields of study. In electronics, it plays a crucial role in the operation of devices such as tunneling diodes and flash memory. Quantum tunneling enables electrons to tunnel through thin insulating barriers, facilitating the flow of current and enabling the miniaturization of electronic components. Without quantum tunneling, the advancements in modern technology that we enjoy today would not have been possible.

Chemistry also benefits from the phenomenon of quantum tunneling. It allows for reactions to occur at lower energies than classically predicted, leading to the discovery of novel chemical reactions and the understanding of enzymatic processes. Quantum tunneling influences the rates of chemical reactions by enabling particles to tunnel through energy barriers, altering the dynamics and kinetics of chemical transformations.

Furthermore, solid-state physics explores quantum tunneling in the context of superconductivity. In superconductors, pairs of electrons known as Cooper pairs tunnel through the energy gap, enabling the flow of electrical current without resistance. This remarkable phenomenon, made possible by quantum tunneling, has significant implications for applications such as magnetic resonance imaging (MRI) and particle accelerators.

The phenomenon of quantum tunneling raises profound questions about determinism and the nature of reality. It challenges our classical notions of cause and effect, highlighting the probabilistic nature of the quantum world. Quantum tunneling invites us to contemplate the intricate dance of particles, the delicate balance between observation and reality, and the interplay of probabilities that governs the quantum realm.

As we delve deeper into the mysteries of quantum tunneling, we uncover not only its practical applications but also its beauty and elegance. It offers a glimpse into the profound interconnectedness of the quantum world, where particles transcend barriers and weave a tapestry of possibilities. Quantum tunneling serves as a reminder that our universe is far more intricate and enigmatic than we can fathom, beckoning us to explore, question further, and push the boundaries of our understanding.

In summary, quantum tunneling is a captivating phenomenon that challenges our classical notions of physical boundaries. Through the wave-like behavior of particles and the probabilistic nature of quantum systems, particles can penetrate energy barriers that would be classically impenetrable. This concept of tunneling finds applications in various fields, revolutionizing electronics, chemistry, and solid-state physics. It enables the miniaturization of electronic devices, uncovers new chemical reactions, and paves the way for superconductivity. Quantum tunneling also poses philosophical questions about determinism and the fundamental nature of reality.

As we conclude our exploration of quantum tunneling, it is evident that this phenomenon holds immense significance in the intricate tapestry of quantum physics. Its beauty lies not only in its practical applications but also in the profound insights it provides into the nature of our universe. Quantum tunneling invites us to embrace the mysteries, challenge our intuitions, and embark on a journey of discovery that continues to unravel the enigmatic secrets of the quantum realm.

Let us carry forward the wonder and curiosity ignited by quantum tunneling as we delve deeper into the captivating world of quantum physics, where extraordinary phenomena and groundbreaking theories await our exploration.

Bonus 2

Physics and Superheroes Download: Unleashing the Science Behind Superpowers. Exploring the Physics of Superhero Abilities

Superheroes have captured our imaginations for decades with their extraordinary abilities and larger-than-life adventures. But have you ever wondered if their powers could actually exist in the realm of science? In this bonus section, we'll delve into the physics behind some iconic superhero abilities, examining the principles that could potentially explain their superhuman feats.

Super Strength: The Marvelous Mechanics

Superheroes like Superman and the Hulk possess immense strength, enabling them to lift heavy objects and overpower their enemies. The question is, how could such incredible strength be possible? The answer lies in the mechanics of the human body and the concept of muscle power. Our muscles contract through the interaction of proteins, allowing us to generate force. However, the strength exhibited by superheroes surpasses the limits of human physiology. Theoretical explanations involve the idea of enhanced muscle fiber density, advanced energy storage mechanisms, or even harnessing the power of exotic particles. While these concepts remain purely speculative, they showcase the potential scientific foundations for super strength.

QUANTUM PHYSICS FOR BEGINNERS

Flight: The Aeronautical Enigma

Characters like Superman and Iron Man can defy gravity and take to the skies, soaring through the air with ease. But how could flight be possible without the aid of aircraft or wings? One possible explanation is through the manipulation of gravitational forces. In the realm of physics, antigravity or gravitational control remains theoretical, but it offers a plausible mechanism for superheroes to overcome Earth's gravitational pull and achieve flight. Another concept involves the use of propulsion systems, such as repulsor technology or jetpacks, to provide the necessary thrust. While current scientific knowledge has not unlocked the secrets of human flight, exploring these possibilities fuels our fascination with the idea of taking to the skies like our favorite superheroes.

Invisibility: The Optical Illusion

The power of invisibility, as possessed by characters like the Invisible Woman or the Predator, presents an intriguing challenge to our understanding of optics and light. Invisibility could be achieved through the manipulation of light waves, bending or redirecting them around the object to render it transparent. The phenomenon of metamaterials, artificially engineered materials with unique properties, holds promise in realizing the concept of invisibility by controlling the path of light. By creating structures that can manipulate electromagnetic waves, invisibility cloaks could potentially be developed. However, the current state of research in this field is still in its infancy, and the practical realization of invisibility remains an exciting but distant possibility.

Time Travel: The Temporal Conundrum

Time travel is a staple of many superhero stories, with characters like the Flash or Doctor Strange bending the fabric of time to their will. However, time travel remains a highly speculative concept in the realm of physics. It involves traversing the fabric of space-time, altering its geometry to navigate through different temporal coordinates. Theoretical frameworks such as wormholes, cosmic strings, or the manipulation of black holes have been proposed as potential mechanisms for time travel. However, the challenges involved, including the immense energy requirements and the preservation of causality, present significant hurdles that are yet to be overcome. While time travel remains firmly in the realm of science fiction, exploring the theoretical possibilities continues to captivate our imagination.

Energy Projection: The Power of Physics

Characters like Cyclops from the X-Men possess the ability to project energy beams from their eyes, showcasing a form of energy manipulation beyond our current scientific understanding. The concept of harnessing and directing energy, however, finds its roots in fundamental physics. It draws inspiration from electromagnetic fields, particle accelerators, and plasma physics. The idea of focusing and releasing energy in a controlled manner could involve advanced technologies or even tapping into the quantum realm. Quantum physics offers insights into the behavior of particles at the smallest scales, where the manipulation and release of energy could potentially be harnessed in extraordinary ways. By harnessing quantum effects or developing advanced energy projection devices, the concept of projecting energy beams may not be entirely out of reach. However, it remains an area of active research and exploration, pushing the boundaries of our understanding of physics.

Teleportation: The Quantum Leap

QUANTUM PHYSICS FOR BEGINNERS

The ability to teleport, as seen in characters like Nightcrawler or the Star Trek teleportation technology, challenges our understanding of space and matter. Teleportation involves instantaneously transferring an object from one location to another without physically traversing the space in between. While teleportation remains speculative, concepts rooted in quantum mechanics offer intriguing possibilities. Quantum entanglement, a phenomenon where particles become inseparably linked across vast distances, could provide a foundation for teleportation. By entangling particles at different locations and manipulating their quantum states, it might be possible to transmit information or recreate an object's state in another location. However, the challenges of maintaining coherence, overcoming the limitations of entanglement distance, and reconstructing complex objects remain significant obstacles to realizing practical teleportation.

In conclusion, the exploration of the physics behind superhero abilities bridges the gap between science and fiction, inspiring us to envision what may be possible in the future. While the concepts discussed in this bonus section are firmly rooted in speculation and the realms of scientific imagination, they provide a glimpse into the potential of scientific discovery and technological advancements. The principles of physics, from mechanics and optics to quantum phenomena, offer a rich playground for exploring the boundaries of human potential. As our understanding of the universe continues to expand, who knows what extraordinary abilities we may one day unlock and what new superheroes may emerge in the realm of science.

Note: The content presented in this bonus section is intended for entertainment and speculative purposes only. It is important to distinguish between scientific fact and fictional narratives, recognizing that the physics behind superhero abilities often involve artistic license and creative storytelling.

Bonus 3

The Future of Quantum Physics: From Discoveries to Breakthroughs. Anticipating the Exciting Frontiers of Quantum Research

Quantum physics has already revolutionized our understanding of the fundamental laws that govern the universe. From its early days of exploring the behavior of subatomic particles to the development of quantum computing and quantum cryptography, the field has made remarkable progress. However, the journey of quantum physics is far from over, and the future holds even more exciting possibilities.

Quantum Computing and Information Processing:

One of the most promising frontiers of quantum research lies in the development of quantum computers. Quantum computers have the potential to perform calculations exponentially faster than classical computers by harnessing the power of quantum bits, or qubits. Scientists and engineers are actively working towards building more stable and scalable qubits, improving quantum error correction techniques, and exploring new quantum algorithms. The realization of practical quantum computers could revolutionize fields such as cryptography, optimization, and drug discovery, unlocking new possibilities and solving complex problems that are currently beyond the reach of classical computers.

Quantum Sensing and Metrology:

Quantum technologies also have the potential to revolutionize sensing and metrology. Quantum sensors, based on the principles of quantum mechanics, can achieve unprecedented levels of precision and sensitivity. For example, quantum gyroscopes can provide highly accurate measurements of rotation, enabling advancements in navigation systems and inertial sensing. Quantum magnetometers can detect tiny magnetic fields, leading to advancements in medical imaging, mineral exploration, and materials characterization. By pushing the boundaries of quantum sensing, scientists aim to develop more efficient and reliable tools for scientific research, industrial applications, and everyday life.

Quantum Communication and Networking:

Quantum communication and networking hold great promise for secure and efficient information transfer. Quantum cryptography, based on the principles of quantum mechanics, enables secure communication by leveraging the properties of quantum entanglement and quantum key distribution. Quantum networks, connecting distant nodes through entangled particles, could enable secure communication channels resistant to eavesdropping and hacking. Researchers are actively working on extending the range of quantum communication, improving the efficiency of quantum repeaters, and developing quantum internet architectures. The realization of a quantum internet could transform how we communicate, ensuring privacy and security in an increasingly interconnected world.

Quantum Materials and Engineering:

Advancements in quantum materials and engineering are opening up new avenues for technological breakthroughs. Quantum materials, such as topological insulators and superconductors, exhibit unique electronic properties that can be harnessed for applications in electronics, energy storage, and quantum computing. Researchers are exploring the synthesis and characterization of novel quantum materials, as well as developing techniques for controlling their properties and integrating them into devices. By understanding and manipulating the quantum behavior of materials, scientists aim to unlock new functionalities and pave the way for innovative technologies.

Quantum Biology and Life Sciences:

Quantum physics is also finding its way into the realm of biology and life sciences. Researchers are investigating quantum phenomena and quantum coherence in biological systems, such as photosynthesis and bird navigation, to understand how quantum effects might play a role in these processes. By unraveling the quantum secrets of life, scientists hope to gain new insights into biological processes and potentially design biomimetic technologies inspired by nature's quantum tricks.

The future of quantum physics is filled with exciting possibilities. As researchers delve deeper into the quantum realm, we can anticipate breakthroughs that will reshape technology, communication, and our understanding of the universe. Collaboration between physicists, engineers, and other scientific disciplines will be crucial in driving progress and unlocking the full potential of quantum physics. The challenges ahead, from maintaining quantum coherence to scaling up quantum technologies, are significant. However, the rewards are equally compelling, with the promise of revolutionary discoveries and transformative advancements that will impact our world in ways we can only imagine.

Quantum Simulation and Modeling:

Another exciting frontier in quantum research is the field of quantum simulation and modeling. Quantum simulators, whether realized in physical systems or through advanced computational algorithms, offer a unique opportunity to study and understand complex quantum phenomena that are otherwise difficult to observe or simulate. By simulating quantum systems, scientists can gain insights into materials properties, chemical reactions, and even biological processes at a level of detail previously unattainable. This capability opens up new avenues for designing advanced materials, optimizing chemical processes, and exploring the behavior of quantum systems under various conditions.

Quantum Machine Learning:

The intersection of quantum physics and machine learning holds tremendous potential for solving complex problems and enhancing artificial intelligence. Quantum machine learning aims to leverage the power of quantum computers and quantum algorithms to accelerate the training and inference processes of machine learning models. Quantum machine learning algorithms can harness the inherent parallelism and computational capabilities of quantum systems to tackle large-scale optimization problems and process vast amounts of data. This field is still in its early stages, but its potential for revolutionizing the field of artificial intelligence is highly anticipated.

Quantum Ethics and Social Implications:

As quantum technologies continue to advance, it is important to consider the ethical and social implications they may bring. Quantum computing, quantum communication, and other quantum applications have the potential to disrupt various sectors and reshape societal dynamics. Issues such as data security and privacy, the impact on employment and workforce, and the accessibility and distribution of quantum technologies need to be carefully addressed. Ethical frameworks and policies should be developed to ensure the responsible and equitable deployment of quantum technologies for the benefit of all.

In conclusion, the future of quantum physics is a captivating journey that holds immense potential for scientific discoveries and technological advancements. From quantum computing and communication to sensing, materials, and even the intersection with other scientific disciplines, the horizons of quantum research continue to expand. As we delve deeper into the mysteries of the quantum world, collaboration between researchers, engineers, policymakers, and society at large will be essential. Together, we can navigate the challenges, seize the opportunities, and unlock the transformative power of quantum physics for the betterment of our world. The quantum revolution has only just begun, and the possibilities are limited only by our imagination and determination to explore the unknown.

Made in United States
Troutdale, OR
10/29/2023